Microsoft® Project Do's and Don'ts
2nd Edition

Sam Huffman, PMP

With contribution by PPM Expert,
Erik van Hurck

MPUG

Copyright © 2019 MPUG

All rights reserved. Absolutely no portion of this publication may be reproduced in any form or by any means, including but not limited to retrieval, transmission or photocopying systems; nor electronic, mechanical, recording or storage systems, without prior written permission of the copyright owner.

Microsoft, Microsoft Windows, Microsoft Project, Project Server, Project Online, Microsoft Office, Microsoft Excel, Microsoft Word, Microsoft PowerPoint and associated logos are trademarks of the Microsoft group of companies. *Microsoft Project Do's and Don'ts* is an independent publication and is neither affiliated with, nor authorized, sponsored or approved by Microsoft Corporation.

All other trademarks are the property of their respective owners.

The information contained in this publication is provided "as is," without warranty. Every effort has been made to ensure the accuracy of information herein. The author and publisher have no liability to any person or legal entity with respect to loss or damages from information contained in *Microsoft Project Do's and Don'ts*.

For general information about MPUG's products, please contact us at www.mpug.com, +1 (734) 741-0841, info@mpug.com or by writing to MPUG World Headquarters, 1143 Northern Blvd #315, South Abington Township, PA 18411 USA.

Acknowledgements

It is upon becoming an author that I realized how important a team is to successfully publish a book. The individuals involved in producing *Microsoft Project Do's and Don'ts* composed the most talented and collaborative team of professionals with whom I have ever worked. To this group, I offer my heartfelt thanks:

Melanie Cosklo, President of MPUG, the users group for Microsoft Project, for sharing my vision and the need for *Microsoft Project Do's and Don'ts*.

Karissa Clampit, Project Manager, Virtual Assistant and Designer, for all the amazing work coordinating, promoting and designing *Microsoft Project Do's and Don'ts*.

Editors, Dian Schaffhauser, for tirelessly helping keep the content brisk and focused as we moved the project from idea to production, and Jana Phillips, for keeping me consistent and on point.

Erik van Hurck for sharing my vision and striving to ensure technical details were correct and clear, and for his thorough examination of the Project Online Desktop Client topic in Chapter 11.

Kyle Brownell, Product Manager, for the creative wizardry behind my videos, recordings, webinars and blog entries and now supporting the intricacies of publishing.

Patricia Huffman, my wife, partner and head cheerleader for decades, and whose grounded observations motivated me to share my thoughts online, in blogs, webinars, whitepapers and now *Microsoft Project Do's and Don'ts*.

About the Authors

Sam Huffman has delivered training internationally since 1990. He gained insight into Microsoft Project while working at Microsoft as a member of the Microsoft Project development and support team. He maintains his intimate knowledge of Project with each new release and is considered a leading authority on the tool, including the newest enterprise features of all versions of Project, Project Server and Project Online. Sam has honed his instructional skills by delivering programs for thousands of people every year.

His reputation for delivering practical wisdom with an upbeat style has established him as one of the world's leading instructors in Microsoft Project. He is a certified Project Management Professional (PMP) and has been a popular contributor to MPUG's webinars and articles. Catch Sam's blog at http://winprojblog.blogspot.com.

Erik van Hurck, author of the newest chapter in this edition, Agile Project Management using Microsoft Project Online Desktop Client, is a Senior PPM consultant for Projectum, a western European Microsoft Partner with offices in Denmark and The Netherlands. Erik is also a Microsoft MVP. As such, Erik assists enterprise customers in adopting the Project Online cloud solution and all Microsoft technology related to Project and Portfolio Management. Erik is a well-established internet authority on these subjects, authoring a personal blog and YouTube channel, as well as frequently contributing to MPUG. Catch Erik's blog at https://www.theprojectcornerblog.com or his YouTube channel at http://bit.ly/YT_TPC.

Table of Contents

Acknowledgements	iii
About the Authors	v
Preface	ix
From the Author	xi
Chapter 1: Harness the Interface	1
Chapter 2: Set Up for Success	21
Chapter 3: Organize the Task List	37
Chapter 4: Enter Task Durations	47
Chapter 5: Sequence Tasks in the Task List	57
Chapter 6: Create Resources and Assign them to Tasks	67
Chapter 7: Baseline Your Project	95
Chapter 8: Track the Project	107
Chapter 9: Reporting	119
Chapter 10: Using Microsoft Project on Agile Based Projects	135
Chapter 11: Agile Project Management using Microsoft Project Online Desktop Client	151
Appendix A: The Organizer	175
Appendix B: Using Microsoft Project Help	179
Appendix C: Steps to Create a Project File	185
Index	187

Preface

Who should read this book

Microsoft® Project Do's and Don'ts is founded and focused in best practices. These include the discipline of project management, as well as the usage of Microsoft Project. You should read it if you are:

- *A Microsoft Project user.* There are many project management techniques and methods supported in Microsoft Project. *Microsoft Project Do's and Don'ts* will help you pick and choose methods and features that save you time and help you get the most out of your investment in this fabulous tool.

- *A project manager only occasionally.* Most project managers have a job or organizational role outside of their project. Juggling both jobs demands a systematic and repeatable approach for success. If you don't know what you don't know—or can't remember between projects—this book will help you identify and prioritize what you need to remember and do to use Project.

- *A novice to project management.* Project management techniques can be hard to model in software. I have approached this problem with simple definitions, easy examples and explanatory screenshots.

- *A Microsoft® Excel user. Microsoft Project Do's and Don'ts* will be your best friend and guide as you learn Microsoft Project. It was designed with your needs in mind from the very beginning.

Still not sure? Watch a free video on this topic now at www.mpug.com/do.

Note to Microsoft Project Server and Project Online Users

Project Server and Project Online are enterprise project management tools. They are driven by business rules and processes and are managed by an administrator. Your administrator should be consulted about how your projects will relate to all of the other projects in your organization. At a minimum, you will need an account and account logon permission to save or publish your project. There may also be rules on governance, preferred methods of tracking and other organizational preferences that will affect your work as you build, save and manage projects using Project.

Conventions Used

I have tried to be specific in the text, descriptions and screenshots. The exact terminology used by Microsoft Project is included wherever possible.

When you see "Do This" or "Don't Do This" in the text and at the end of each chapter, it's time to pay attention. These are the focal points of the chapter!

 Do This: Contains text that explains a feature, will save you time or further covers a concept.

 Don't Do This: Contains steps to avoid errors and frustration.

Versions of Project Covered

The versions of Microsoft Project used in this book are limited to Project Professional 2013 and higher.

Want more of Sam's Do's and Don'ts? You'll find those—along with other free Microsoft Project resources—at www.mpug.com/do.

From the Author

I created this book for only one reason: to save Project users from the problems brought about by not following best practices. Best practices enable the tool to better represent your project and its pressing realities in time, work, cost and quality. The list of best practices is not that long nor is it overly complicated. It is the result of many years of learning and usage of Project on my part sprinkled with the *occasional* mistake. Learning what to do and not do in Project has been an integral part of my career, and I hope to save you time and frustration in yours.

I encourage you to read the book in sequence from cover to cover. The Do's and Don'ts start before you enter your first task and continue through the life of the project and the Microsoft Project file. If you jump around within a given chapter, you might miss an important fact or technique that's worth knowing.

Enjoy the book and good luck in your projects!

—*Sam Huffman, Bothell, WA*

Chapter 1: Harness the Interface

In this chapter, you will learn:

- The components of the interface and how they interrelate
- How to create and use command shortcuts to save time and confusion

Structure of the Interface

Microsoft Project's interface is similar to other Microsoft Office programs. From the ever-present Quick Access Toolbar to commands and tabs in the Ribbon, the consistent Office look and feel will give you the confidence to continue. There are differences in the content of Project's Quick Access Toolbar and Ribbon, but the concept is the same as in Microsoft Office: a fast, discoverable interface.

The Quick Access Toolbar, the Ribbon, tabs, command groups and Views (specific ways of looking at a project) are the major components controlling the interface. Let's look at each in turn:

Figure 1.1: The Ribbon and Quick Access Toolbar.

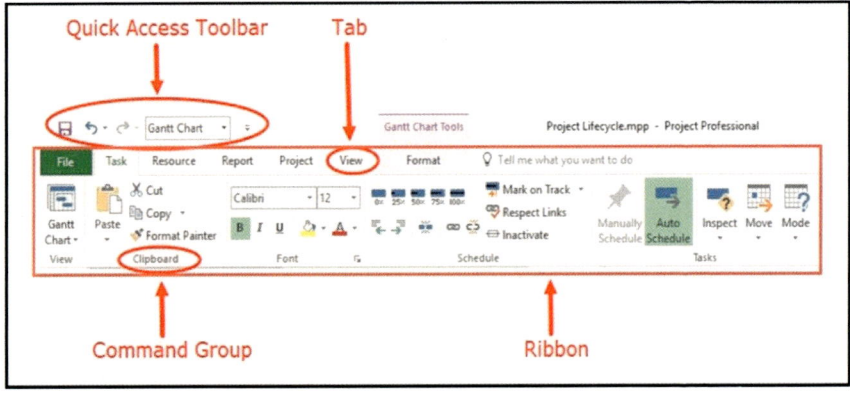

Quick Access Toolbar

The Quick Access Toolbar exists to save you time by placing frequently used commands at the top of the screen. Commands placed in the Quick Access Toolbar are always available and never hidden. This visibility gives the toolbar strategic importance for you in specialized operations. Inserting a frequently used command in Quick Access means not having to search the Ribbon for it anymore. And that saves you time.

The Quick Access Toolbar is just above the Ribbon and to the far left, where commands can be quickly added or removed to meet your needs. If a command exists in the Ribbon, you may make it visible in the Quick Access Toolbar by right-clicking on the command's button and choosing "Add to Quick Access Toolbar."

If the command is not in the Ribbon, you can add it to the Quick Access Toolbar by choosing "More Commands…" in the dropdown just to the right of the Quick Access Toolbar. This will enable the "Customize Quick Access Toolbar" dialog found in Options.

Figure 1.2: Customizing the Quick Access Toolbar.

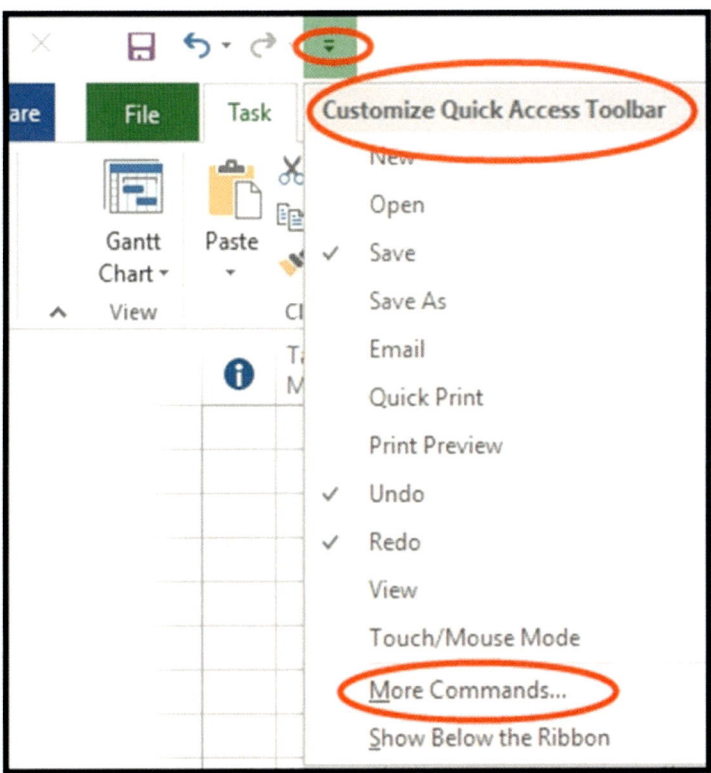

Do This: You will want "quick" access to your most frequently used commands. When you find a command that you need to access and don't want to bounce around between the tabs, add it to the Quick Access Toolbar!

 Don't Do This: Don't forget that Project integrates with Office and shares a very similar interface. If you struggle with finding useful commands in other Office applications, try modifying the application's Quick Access toolbar to suit your needs and to save time!

The Ribbon

The Ribbon is the most prominent feature of the interface. It is context sensitive, which means it will change its content depending on what View is displayed. The context sensitivity will announce its presence as a "Tools" title near the Format tab. If you change Views, you will have View-specific tools and formats available.

Tabs

Unlike the Quick Access Toolbar, the Ribbon organizes commands in tabs along the top edge and then in command groups along the bottom edge. Tabs are categorical. Command groups are functional. For example, the Task tab offers command groups to organize and schedule your project's tasks.

Figure 1.3: Tabs are major components of the Ribbon.

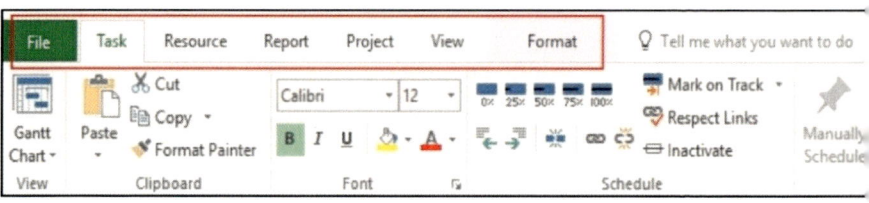

Out of the box Project is configured to show you seven tabs. Each tab has a different purpose and thus different content than the others. The content of the tabs is also dependent upon the View that is applied. Here is each tab and its purpose:

File tab: Its name tells all. This tab is your primary control tool for files. Use it to create new files, save, import and export in other file formats, change the file name, print and collaborate. Lots of Project users call this tab and its many functions the "Back Office."

Task tab: Use this tab when creating, organizing and sequencing tasks within your project and estimating variables such as the amount of time and work required to complete a task. The Task tab is also very useful in troubleshooting and tracking schedules.

Resource tab: This tab assists in defining project resources (such as people), assigning them to tasks, refining their schedule and resolving problems in assignments.

Report tab: New to Project 2013, this tab contains Visual Reports and View Reports. Visual Reports export project metadata to Microsoft Excel or Microsoft Visio for charting purposes. View Reports takes the project data and charts the data without having to export it. View Reports can be formatted and saved. The report may also be customized, saved and made available for future projects.

Project tab: Project-level operations are set from this tab. It is from here that you can tell Project which days and hours of the week are working times, set the starting date of the project and prepare the file for tracking schedule performance to the plan and report on project progress.

View tab: This tab provides flexibility in viewing and analyzing your project information from different perspectives. It can help you switch between topics of interest quickly, look for project information categorically and conduct other operations such as sorting and filtering data.

Format tab: Data formatting in Project relates to the state of tasks, resources and schedules. For example, if a resource is assigned more work than his or her schedule can accommodate, the format of the font identifying the resource will be bold and in red. The Format tab provides the ability to show or hide the amount of formatting required. The content of this tab is context sensitive and extremely dependent on the View you have applied at the moment.

Command Groups

As I mentioned earlier, command groups are subdivisions of the Ribbon and tabs that organize buttons and commands functionally. The content of any command group may vary and be active or grayed out depending on the view applied.

Figure 1.4: Command Groups are major components of each tab.

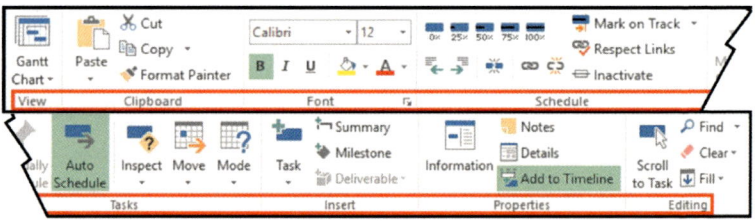

Views

Views are the window into your project data. They are the key players in all aspects of Project. A view is a combination of sheets, tables containing fields, charts, graphs and forms. Additionally, views may be full screen or split into two different windows.

 Do This: Add the "View" list into the Quick Access Toolbar. Click the Quick Access Toolbar's dropdown on the toolbar's right edge and then click on "View" in the list. This will insert and activate a list of the views available in Project. Project's views are now always available for you.

Don't Do This: Don't leave commands that you seldom use in the Quick Access toolbar. Remove them to make room for those that you use frequently!

Figure 1.5: Customizing Quick Access to show views.

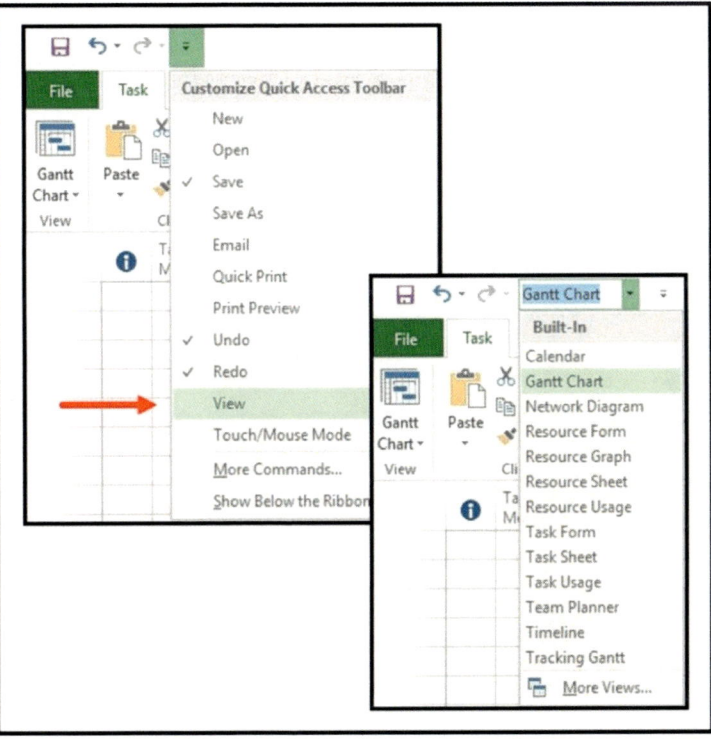

Ch 1: Harness the Interface | 7

Figure 1.6: A Gantt chart showing view components.

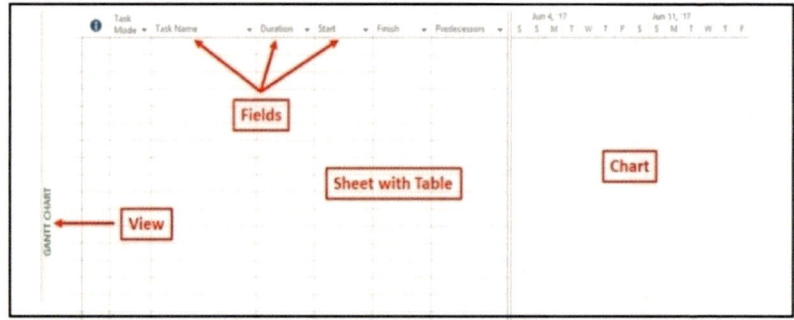

Full Screen View: Many views can be shown in full screen to offer general information at a glance. The Gantt chart and network diagram are examples of a single view shown in full screen.

Combination View. A combination view is two views split into horizontal halves on your computer screen. The most common form of this is the task entry view. The top half contains the Gantt chart while the bottom half contains the task form. When selecting a task in the top, the details of the task are shown in the bottom. Combination views are excellent for drilldown analysis.

Figure 1.7: The Task Entry View.

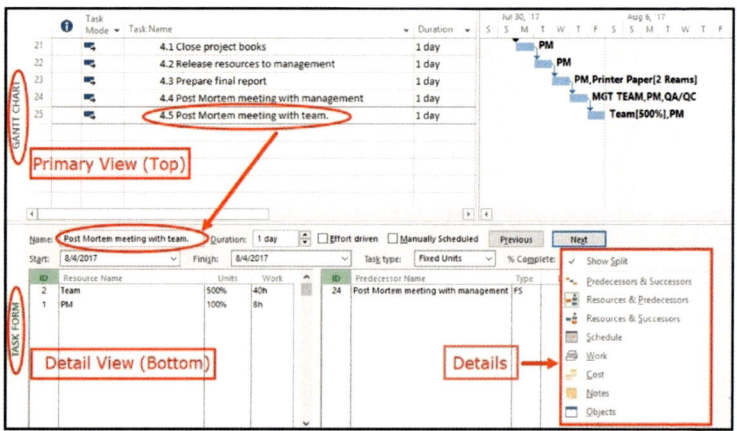

In the Quick Access Toolbar, click on the dropdown in the view list that you created earlier. Now click on the last item in the list: "More Views..." Finally, click on the View "Task Entry." Note that the task form in the bottom pane can show different types of information when you right-click on the form and then choose a different topic. Selecting "Show Split," accessible from the right-click menu, will close the task form.

Figure 1.8: The Task Sheet. This is also the sheet used by the Gantt chart.

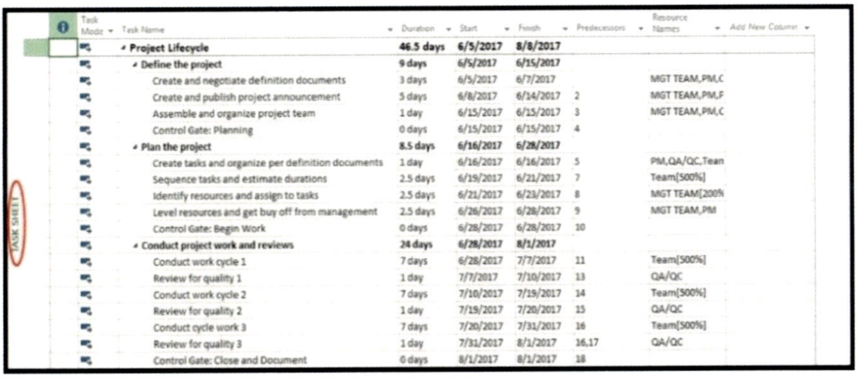

	Task Mode	Task Name	Duration	Start	Finish	Predecessors	Resource Names	Add New Column
		▲ Project Lifecycle	46.5 days	6/5/2017	8/8/2017			
		▲ Define the project	9 days	6/5/2017	6/15/2017			
		Create and negotiate definition documents	3 days	6/5/2017	6/7/2017		MGT TEAM,PM,C	
		Create and publish project announcement	5 days	6/8/2017	6/14/2017	2	MGT TEAM,PM,F	
		Assemble and organize project team	1 day	6/15/2017	6/15/2017	3	MGT TEAM,PM,C	
		Control Gate: Planning	0 days	6/15/2017	6/15/2017	4		
		▲ Plan the project	8.5 days	6/16/2017	6/28/2017			
		Create tasks and organize per definition documents	1 day	6/16/2017	6/16/2017	5	PM,QA/QC,Team	
		Sequence tasks and estimate durations	2.5 days	6/19/2017	6/21/2017	7	Team[500%]	
		Identify resources and assign to tasks	2.5 days	6/21/2017	6/23/2017	8	MGT TEAM[200%]	
		Level resources and get buy off from management	2.5 days	6/26/2017	6/28/2017	9	MGT TEAM,PM	
		Control Gate: Begin Work	0 days	6/28/2017	6/28/2017	10		
		▲ Conduct project work and reviews	24 days	6/28/2017	8/1/2017			
		Conduct work cycle 1	7 days	6/28/2017	7/7/2017	11	Team[500%]	
		Review for quality 1	1 day	7/7/2017	7/10/2017	13	QA/QC	
		Conduct work cycle 2	7 days	7/10/2017	7/19/2017	14	Team[500%]	
		Review for quality 2	1 day	7/19/2017	7/20/2017	15	QA/QC	
		Conduct cycle work 3	7 days	7/20/2017	7/31/2017	16	Team[500%]	
		Review for quality 3	1 day	7/31/2017	8/1/2017	16,17	QA/QC	
		Control Gate: Close and Document	0 days	8/1/2017	8/1/2017	18		

Sheets: A sheet looks a lot like a worksheet in Excel. Closer examination will reveal that this is not the case. The columns in a sheet in Project are actually fields contained in the Project file. The task sheet is a view as well as the left side component in Gantt charts, used for defining tasks. The resource sheet is a view used in defining resources prior to making assignments of tasks in the project.

Tables: Every sheet must have a table to display fields. Only one table can be displayed at a time in a sheet, so it is important for the purposes of navigation and analysis to know which table is currently applied when working in Project. A list of tables can be found by navigating to the View tab, Data command group, and then clicking on the "Tables" button. This will offer you a list of Tables; the current Table will be checked.

Do This: Put the "Tables" button in the Quick Access Toolbar. In the View tab, right click on the "Tables" button and choose "Add to Quick Access Toolbar." Now both views and tables are instantly accessible from anywhere in Project.

Fields: Fields are represented as columns in sheets, as a dropdown item or editable detail in forms and dialog boxes.

There are hundreds of fields in Microsoft Project. Many are customizable, meaning they can contain lookup tables, formulas and graphical indicators. Customizing fields is beyond the scope of this book; however, it is well documented in Project's Help function.

To see a list of fields available for insertion into a Gantt chart sheet, click and hold the mouse cursor on the vertical divider bar between the sheet and the chart. Drag the divider bar to the right and beyond the "Add New Column" field. Click on the dropdown indicator on the right edge of the "Add New Column" field. The list you see contains the fields available for insertion into the sheet.

Figure 1.9: The Duration column is a field.

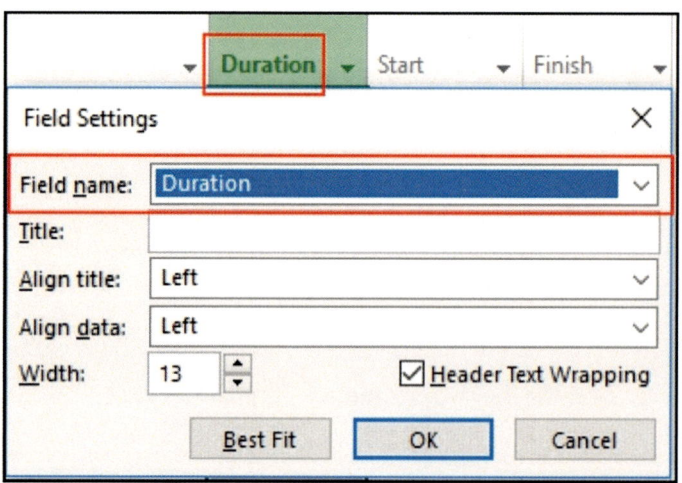

Figure 1.10: A partial list of fields shown after clicking on the "Add New Column" field.

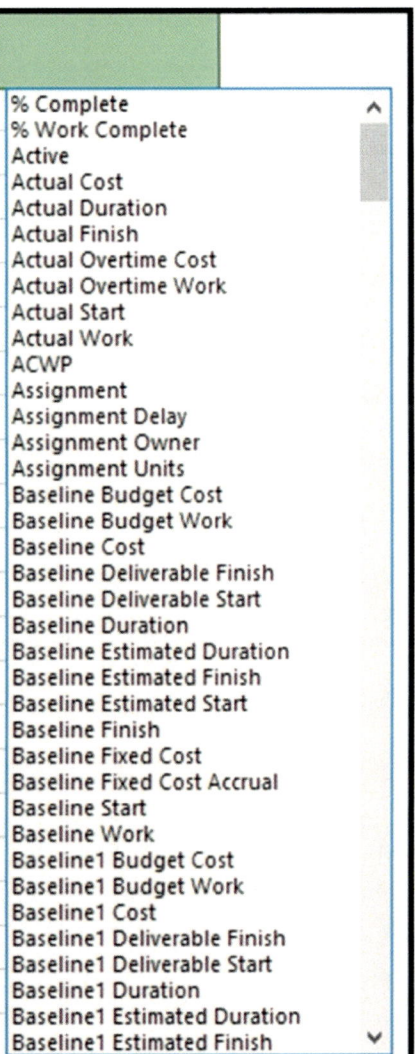

The table for each sheet has its specific fields already created; but should you find the need to show a different field than the default, simply locate it on the list and click on it. Project will insert the field just to the left of the "Add New Column" field.

Figure 1.11: Add New Column.

Finish	Predecessors	Resource Names	Add New Column
8/4/2017			
6/15/2017			
6/7/2017		MGT TEAM,PM,C	
6/14/2017	2	MGT TEAM,PM,P	
6/15/2017	3	MGT TEAM,PM,C	
6/15/2017	4		
6/26/2017			

Do This: Insert a new field. Click on the "Add New Column" dropdown. Note the long list of fields. You don't have to scroll through the whole list to click on the field you want to insert. Press the "w" key on your keyboard and Project will show you only the fields beginning with "w". This is a real timesaver for quick sheet modifications!

Do This: Understand your columns! The titles in columns contain information defining the column and what it does. The fastest way to see this information is to hover the mouse cursor over the column title. Project will display an information box containing its purpose.

Figure 1.12: Hover on a field title to see the field definition.

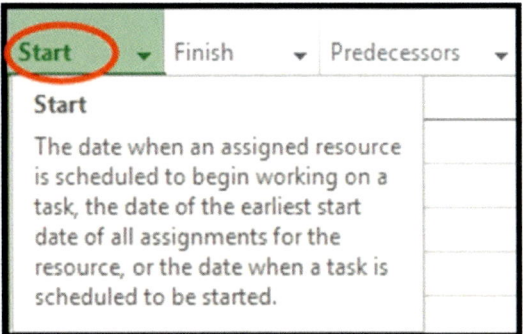

Charts and Graphs: In Project, charts and graphs primarily compare data graphically. The Gantt chart is an example of a chart that compares the length of time expected to do the work in each task and how tasks relate to each other. This type of chart is shown in a timeline or calendar to illustrate task schedule and completion dates.

Figure 1.13: The Gantt Chart option compares task durations and timing.

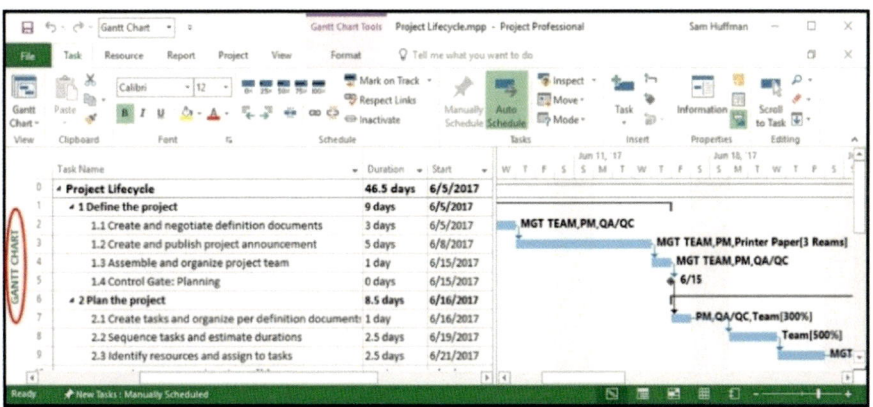

Ch 1: Harness the Interface | 13

The resource graph shows the effect of assigning resources to tasks. Concurrent assignment of tasks can result in over-assignment and unrealistic scheduling. This type of graph can point out when in time the issue exists and help in the analysis needed to resolve the scheduling conflicts.

Figure 1.14: The Resource Graph illustrates how busy a resource is over time.

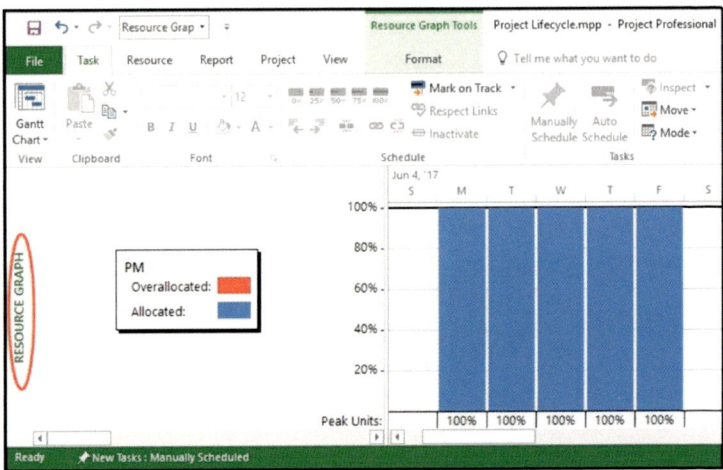

Forms: Usually forms are shown on the bottom of combination views. They show the details of what is specifically selected in the top. Often underused, forms are a superb tool for focusing on specific tasks and resources during their definition and later in the tracking of your project.

Figure 1.15: The Task Form in the bottom of a combination view.

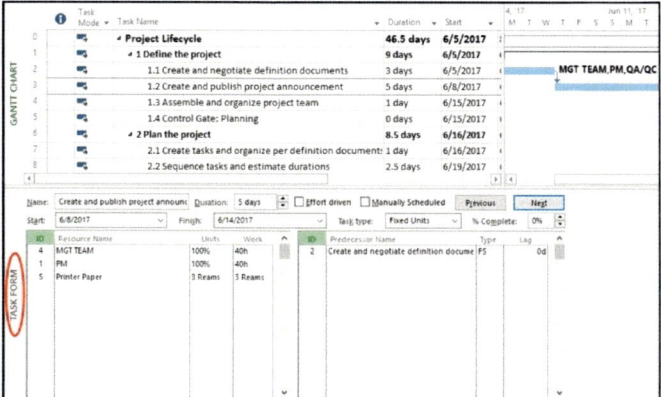

Recommended Views and Their Purpose

Gantt chart: Organizing, sequencing and estimating the duration (length) of tasks in the project.

Tracking Gantt: Tracking performance and rescheduling tasks as needed. This view will also compare the current schedule to the baseline schedule, enabling you to visualize when work is not proceeding quickly enough or starting earlier than planned.

Figure 1.16: The Tracking Gantt compares scheduled task dates and durations to Baseline dates and durations.

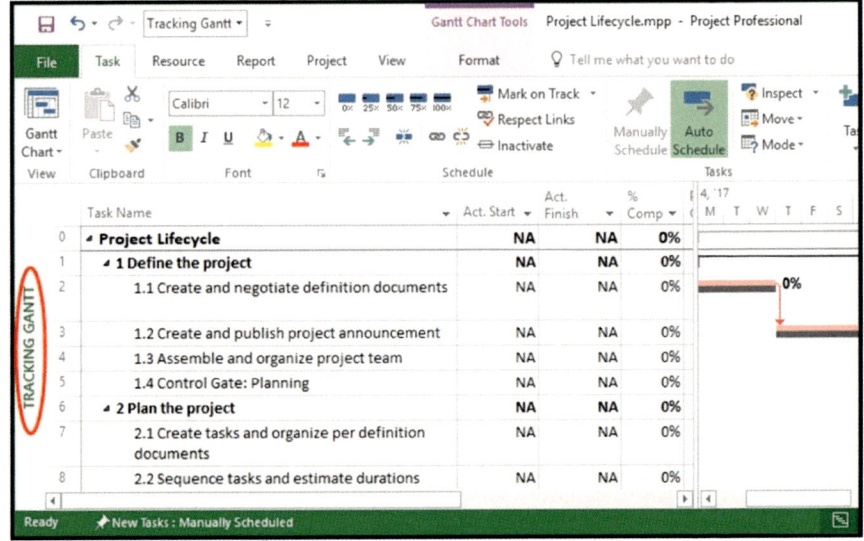

Resource Sheet: **Defining resources (such as people or materials) prior to assignment on a project.**

Figure 1.17: The Resource Sheet provides specific resource information.

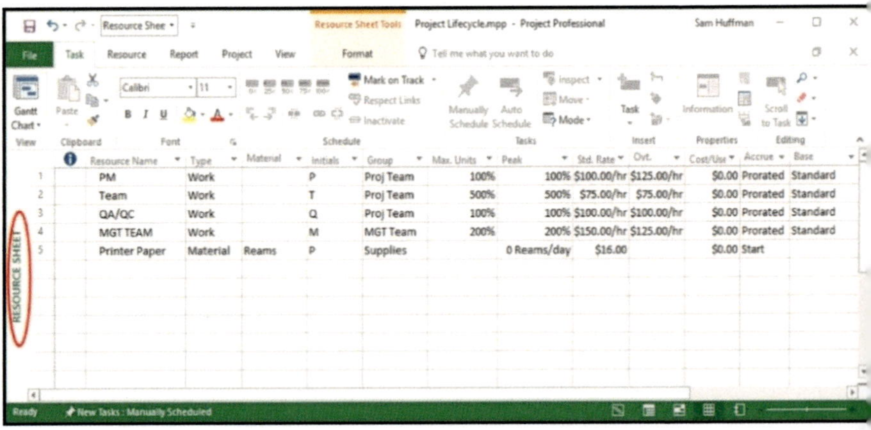

16 | Ch 1: Harness the Interface

Resource Graph: Evaluating the effect of assignments to tasks. If over-assignment occurs, some rescheduling or re-planning may be required.

Figure 1.18: The Resource Graph.

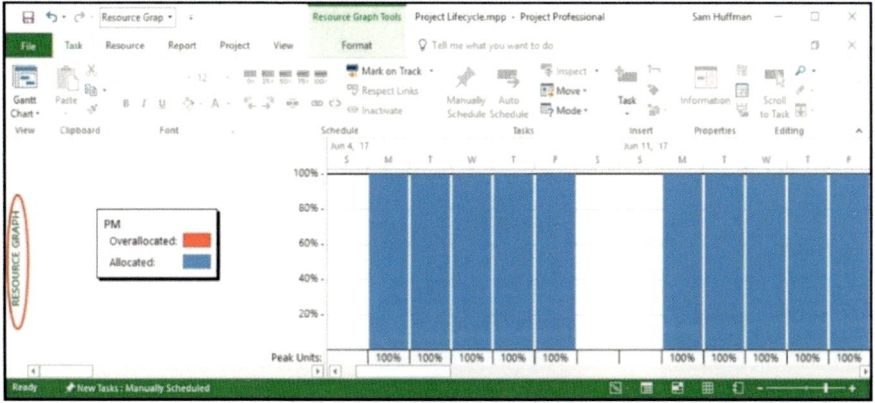

Figure 1.19: The Project Timeline showing the project phases, control gates and dates.

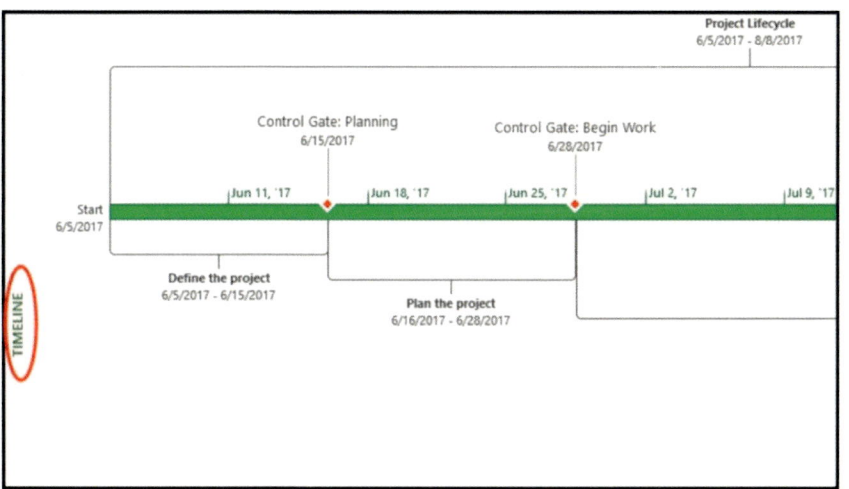

Timeline: Great for those times when you need to see the project "at a glance." Useful also for presentations, email and reporting. The timeline in Project 2016 can have up to 10 different timelines in each timeline view. You can now separate the phases and other elements of your project into different timelines! Having less complex timelines allow you to focus and compare dates more easily and with less confusion. For example, one timeline could show a meeting schedule and another the control gates. You get to show what you want and so does the viewer.

Project has an interface that can be deceptively hard to work with. It is to your advantage to understand the components of it and mold the interface to meet your needs.

Do's and Don'ts Review

Do's:

- Add your most frequently-used commands to the Microsoft Project Quick Access Toolbar.
- Add the "View" list into the Quick Access Toolbar.
- Put the "Tables" button in the Quick Access Toolbar.
- Add a new field to a sheet with the "Add New Column" field.
- Understand your columns. They are fields!

Don'ts:

- Don't forget that Project integrates with Office and shares a very similar interface. If you struggle with finding useful commands in other Office applications, try modifying the application's Quick Access toolbar to suit your needs and to save time!
- Don't leave commands that you seldom use in the Quick Access toolbar. Remove them to make room for those that you use frequently!

Chapter 2: Set Up for Success

In this chapter, you will learn:

- How to set the project start date
- The calendars used by Microsoft Project
- The hierarchy of calendars
- How to create, modify and set calendars
- How to choose the method of scheduling in Project

When creating a project for the first time many project managers begin by typing information in the "Task Name" column. Later they find that Microsoft Project is calculating working time incorrectly. Task dates begin to look inaccurate and the Project duration begins to change. Finally, they may even abandon the use of Project in managing their project. The problems began with the first data entry. Project was not set up correctly for data and so errors ensued.

Project is like other database applications in that the correct set-up and configuration of the software will drive accuracy. In Project that accuracy is critical for calculations necessary to create a realistic schedule.

 Don't Do This: Don't start a new project by entering data! Establish a Project Start Date and applicable Calendar for Project to produce a more realistic task schedule.

Set the Project Start Date

Microsoft Project schedules tasks more accurately when you tell it the starting date of your project. The start date is also important because Project uses that date as a default start date for new tasks and tasks that are not sequenced with other tasks.

 Do This: When creating a new project, always set the project start date. This is especially critical if it started in the past.

The start date does not have to be the current date! In many cases the project actually started before the file is created, and so the real date should be used. In this way, work can be modeled and captured realistically. Once the start date is set, click on "OK." Your file is now ready for calendar selection.

The project start date is set in the Project Information dialog. Navigate to this dialog by selecting the Project tab, then click on the Project Information button in the Properties command group. The start date can be typed into the Start date box or selected from a calendar by clicking on the Date dropdown situated on the right side of the Start date box.

Figure 2.1: Project Information highlighting the project start date and project Calendar selection.

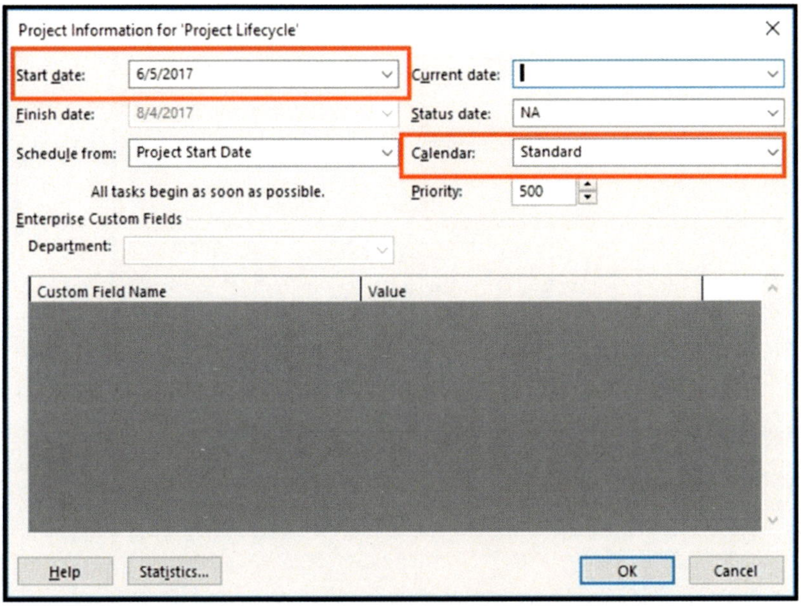

[Project Information dialog box for 'Project Lifecycle' showing Start date: 6/5/2017, Finish date: 8/4/2017, Schedule from: Project Start Date, Current date, Status date: NA, Calendar: Standard, Priority: 500]

The Calendars Used by Project

It seems an easy concept—just use a calendar to schedule your project. It is a complex set of instructions that informs Project of when work can or can't be scheduled for the project, shift, resources and tasks. Without a clear set of rules on this, the project would be a list of conflicting task dates that Project could not schedule.

Calendars are created and modified in the Change Working Time dialog. Navigate to the Change Working Time dialog by selecting the Project tab, then click on the Change Working Time button in the Properties command group.

Figure 2.2: Change Working Time dialog showing calendars and current work time settings.

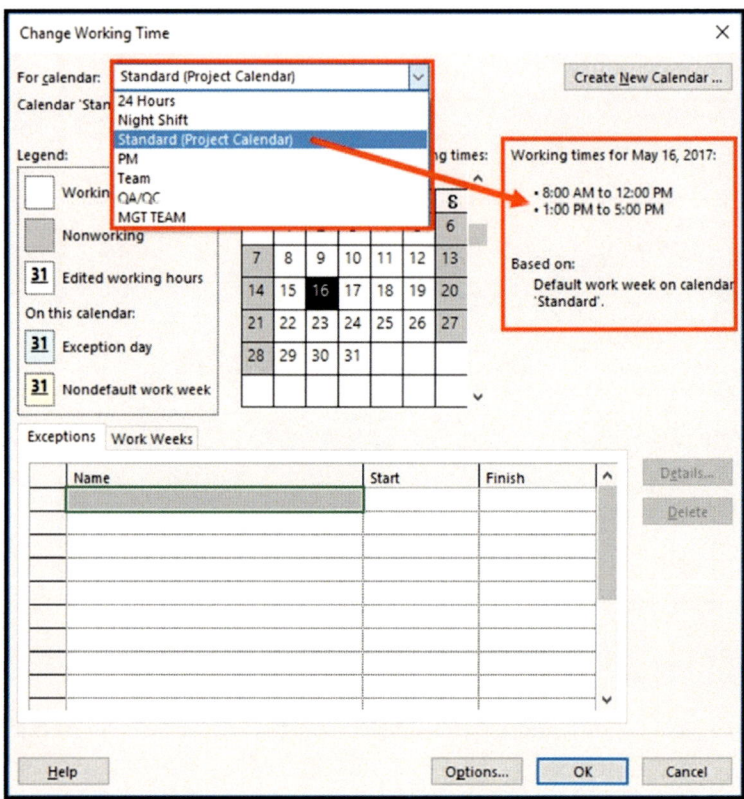

There are four types of calendars used in Microsoft Project:

The Project calendar: Used by Project as the base or default working calendar. It contains the common working dates and times used by all resources assigned to the project. The default project calendar is called "Standard." A project's calendar is set in the Project Information dialog.

Shift calendars: These are custom base or default calendars used by categories of resources instead of the Project Calendar. Examples of a shift calendar are "Part Time," "Weekend" or "4 X 10." They are essentially any calendar used as the foundation for a Resource Calendar and are assigned to resources in the Resource sheet.

Resource calendars: Every working resource in the project has a calendar that identifies resource-specific non-working periods such as vacation dates, sabbaticals or leaves of absence. Resource calendars are created automatically upon creation of the resource in the Resource sheet.

Figure 2.3: Calendar assignments on the Resource sheet.

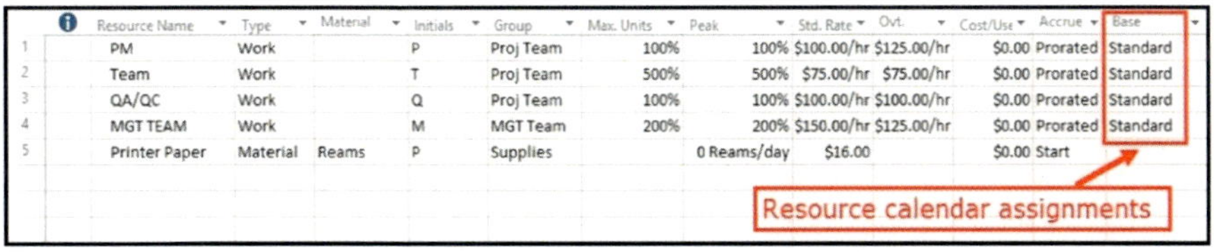

	Resource Name	Type	Material	Initials	Group	Max. Units	Peak	Std. Rate	Ovt.	Cost/Use	Accrue	Base
1	PM	Work		P	Proj Team	100%		100%	$100.00/hr	$125.00/hr	$0.00	Prorated	Standard
2	Team	Work		T	Proj Team	500%		500%	$75.00/hr	$75.00/hr	$0.00	Prorated	Standard
3	QA/QC	Work		Q	Proj Team	100%		100%	$100.00/hr	$100.00/hr	$0.00	Prorated	Standard
4	MGT TEAM	Work		M	MGT Team	200%		200%	$150.00/hr	$125.00/hr	$0.00	Prorated	Standard
5	Printer Paper	Material	Reams	P	Supplies			0 Reams/day	$16.00		$0.00	Start	

Resource calendar assignments

Task calendars: This calendar is rarely used. Task calendars may be used to schedule a task in a period that the other calendars would not. The task calendar can be manually set to have priority over resource calendars in the Task Information dialog. An example of using a Task calendar is any task that has a schedule unique to the task, such as a weekend office move. In this case, the resources assigned to the task would be expected to participate in the move, even if the resource calendar was set for a non-working weekend. You get to the Task Information dialog by double-clicking on the task name in the task list. Navigate to the Advanced Tab on top of this menu to see information related to the task calendar. Microsoft Project remembers the last Tab you visited in the Task Information menu, so look carefully at any tabbed dialog you are offered.

Figure 2.4: The Task calendar noting that resource calendars are ignored for scheduling the task.

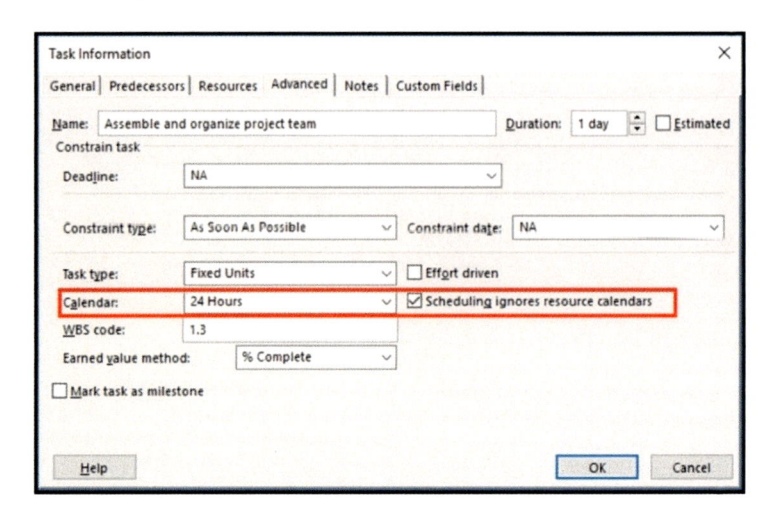

The Hierarchy of Calendars

Your project can have multiple calendars that are mutually exclusive. This begs the question, which calendar wins in a conflict? There is a hierarchy that Project uses in settling calendar conflicts:

1. The Project calendar always loses in a calendar conflict.
2. The Shift calendar wins in any conflict with the Project calendar, but loses to the Resource and Task calendars.
3. The Resource calendar loses only to a Task calendar that is set to override Resource calendars.
4. The Task calendar never loses when it is set to ignore Resource calendars. Otherwise, the Resource calendar and Task calendar are equal in precedence.

Create, Modify and Set Calendars

You do not have to use all calendar types to use Project effectively. In fact, I recommend that you use only the Project and Resource calendars until there is a need for Shift and Task calendars. In support of this, I will confine my comments to only the Project and Resource calendars.

> Do This: Keep calendars simple at first. As your skill level with Project improves, try the other calendar types as required. Many users find that only Project and Resource calendars are needed to create a well-designed project schedule.

Create a new Project Calendar

You may create as many calendars as you wish, but there can be only one project calendar applied to a project file.

To create a new project calendar:

1. Select the Project tab, then select "Change Working Time."
2. Click on "Create New Calendar."
3. Give your calendar a name. You should start with making your calendar a copy of the "Standard" calendar, the default Project calendar.
4. By clicking OK you create the new calendar.
5. To apply the calendar, go to the Project tab, then click on Project Information. Select the new calendar from the list of calendars and OK the dialog. The new calendar will now be used instead of "Standard."

Figure 2.5: Create a new calendar and any working date exceptions in the Change Working Time dialog.

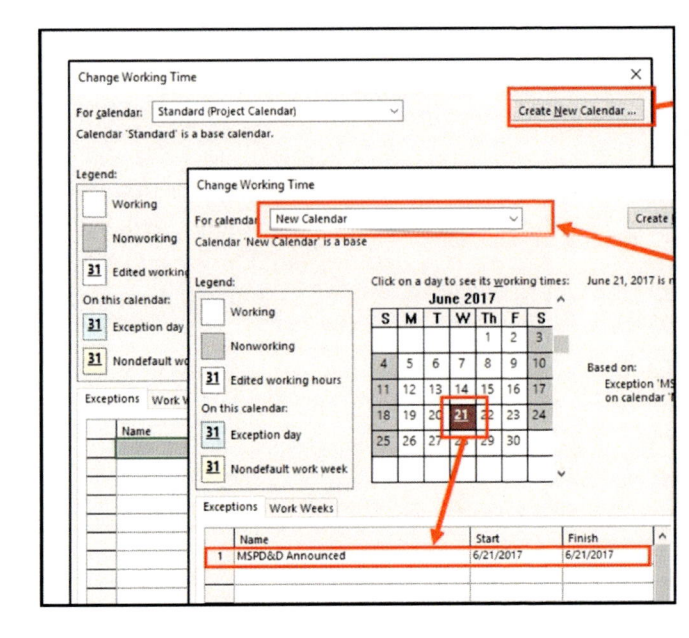

Modify a Project Calendar

When you modify a project calendar, the changes affect every resource calendar that uses the project calendar as its foundation for scheduling. The most common type of modification is the calendar exception. Calendar exceptions are dates that are identified to the system as non-working. Any date that your organization takes off, as a holiday, for example, is an exception that must be noted in the project calendar. You create calendar exceptions manually.

To create a calendar exception:

1. Bring up the calendar you wish to edit in the "Change Working Time" dialog.
2. Find the date or date range you wish to modify in the calendar and select it.

3. Type the title or name of the exception into the first available "Name" cell in the Exceptions list.

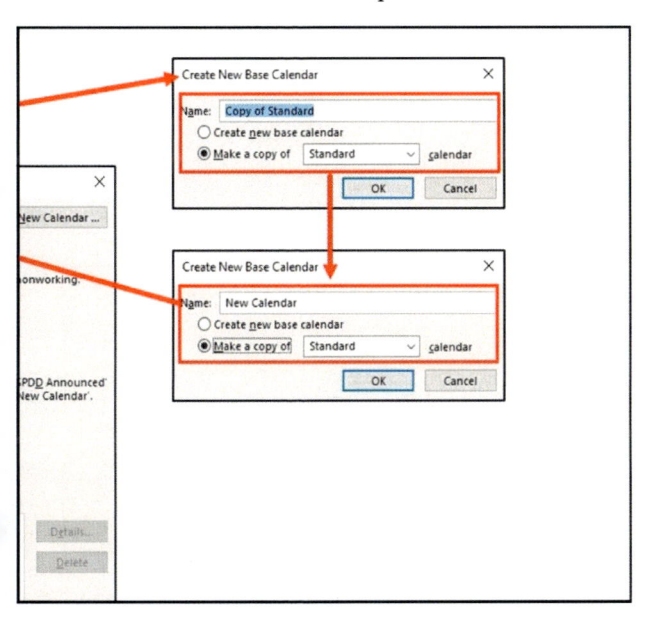

Every calendar may also be edited to ensure the correct working time. The default working times are 8AM to 12PM, then 1PM to 5PM. If your organization starts or ends the workday at a different time and you wish to modify the working time, you must manually adjust it in the "Work Weeks" tab in the "Change Working Time" dialog. The new work time settings will take place immediately.

Steps to Modify Working Time

1. Bring up the calendar you wish to edit in the "Change Working Time" dialog.
2. Click on the "Work Weeks" tab in the lower half of the "Change Working Time" dialog.
3. Click on "Default" in the "Work Weeks" list.
4. Click on the "Details..." button.
5. Select the days of the week you need to modify.

6. Click on "Set day(s) to these specific working times."
7. Enter the working times, including AM or PM.
8. If a nonworking day is to be set, choose "Set days to nonworking time."
9. When entry is complete, click on OK at the bottom of the dialog.

Figure 2.6: Change work times in a calendar.

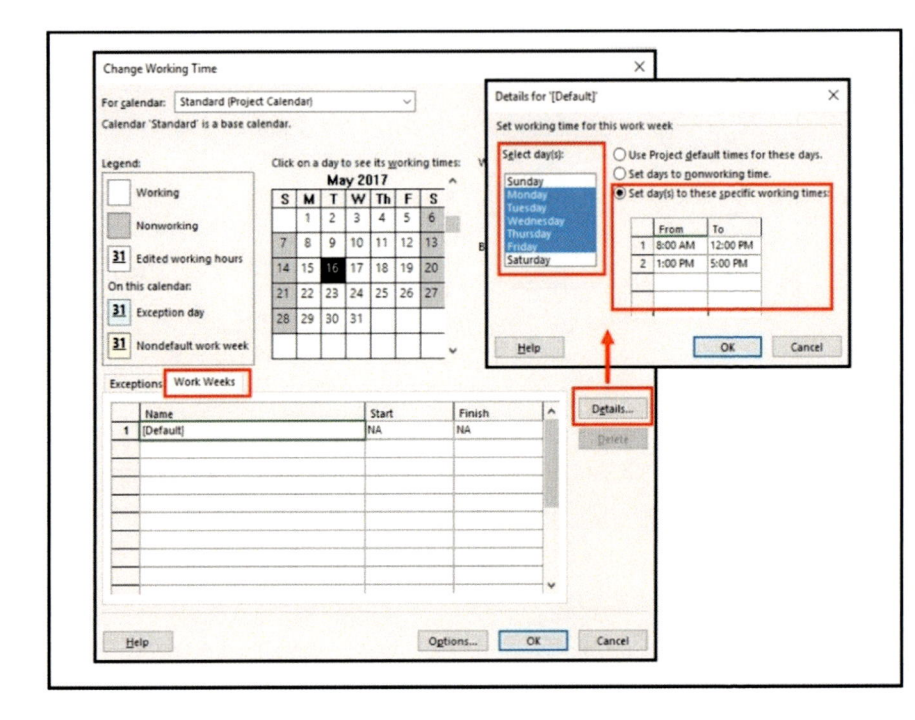

Resource Calendar

A resource calendar is created when a "work" resource is created. A work resource is a person or tool or project material that has a cost rate. When the resource is human, the rate may be by the minute, hour, week, month or year. Once the resource is created, the Resource calendar is then used for "exceptions" to the Project calendar. Exceptions in a resource calendar might include vacations, sabbaticals, maternity and paternity leave or other resource-specific exceptions to what is a working day for other resources. Every resource calendar created will be listed in the "Change Working Time" dialog's calendar list. Resource calendars are edited in the same manner as the project calendar.

> Don't Do This: Don't re-create calendars for every file! Use the Organizer to copy calendars between projects. The Organizer is discussed in the Appendix.

Choose a Method of Scheduling

Microsoft Project can do wondrous things with a schedule—if it is allowed to. There are two methods that Project can use in calculating a schedule: scheduling from the project start date so the schedule can occur as early as possible, and scheduling from the project finish so that the schedule is as late as possible. Both options are available in the Project Information dialog.

Schedule from the project start date: This method is the default, and is the easiest to understand. The starting date of the project is where calculations begin. From that point in time forward, task start and finish dates are calculated by employing task durations and sequencing.

Figure 2.7: Project Information dialog's "Schedule from" list. The choices are from the project start date or project finish date.

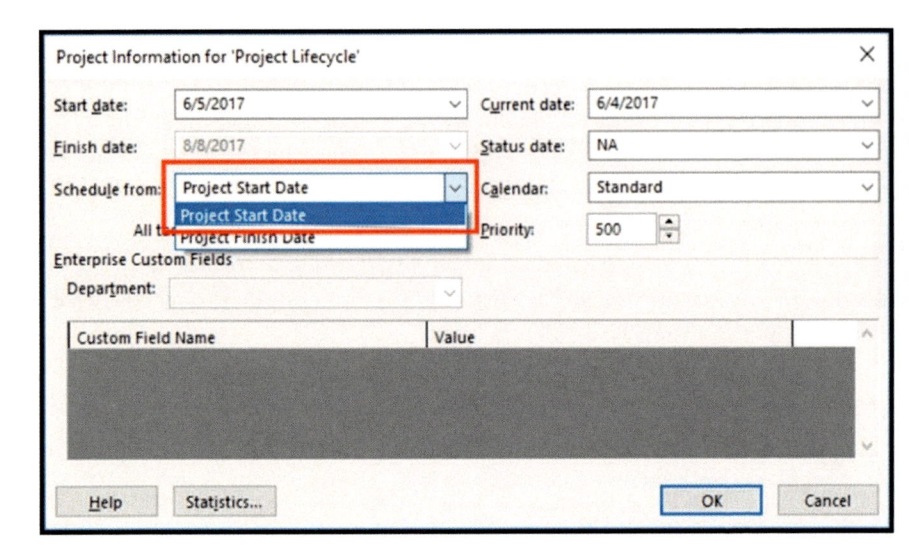

Project Information for 'Project Lifecycle'				✕
Start date:	6/5/2017	Current date:	6/4/2017	
Finish date:	8/8/2017	Status date:	NA	
Schedule from:	**Project Start Date**	Calendar:	Standard	
	Project Start Date	Priority:	500	
All t	Project Finish Date			
Enterprise Custom Fields				
Department:				
Custom Field Name		Value		
Help	Statistics...		OK	Cancel

Schedule from the project finish date: This method of scheduling starts from a date you select as the project finish date, then calculates task start dates by employing task durations and sequencing and working from the end date of the project. Tasks are scheduled as late as possible, so there is no flexibility in the schedule. Some call this "Drop Dead Scheduling."

> ✓ <u>Do This:</u> Schedule from the start date when setting up a project. This method of scheduling allows for schedule flexibility. Scheduling a project from the finish allows for no schedule flexibility because tasks are scheduled as late as possible.

Manual and Automatic Scheduling

In Project 2010 a new scheduling technique was introduced to lower the skill level required to enter data. Manual scheduling is now the default technique used in Project 2010, 2013 and 2016. In all previous versions of Project the only option was to Automatically Schedule a project. They have very different behavior. You can get confused if you do not understand the differences, so let us cover that.

Manual scheduling: Manual scheduling allows for easier and faster project design, but there can be drawbacks when scheduling resources or resolving sequence and work issues. In this scheduling technique Project's scheduling engine is prohibited from changing any schedule data entered. Potential conflicts are formatted and visible, but correction and updating is left to the user and is often very time consuming.

Figure 2.8: Manually Schedule and Auto Schedule buttons. The buttons are on the Task tab in the Tasks command group.

Automatic scheduling: This technique actively watches over all input into the schedule. It will make automatic corrections, flag errors and offer assistance. This technique requires some patience as it can interrupt the flow of entering information, but it is helpful in decreasing the chances of scheduling errors.

Which technique is best for you?

If you are new to scheduling and project management, stay with manual scheduling for a while until you understand Project a little better. Even when you are an experienced user, you will probably mix both techniques in a single Project file.

If you are more experienced or are an advanced user of Project, you may prefer to change the default to Automatic Scheduling immediately. This is accomplished by clicking on "New Tasks:" found in the Status Bar in the lower left of the Project interface. This will offer you the opportunity to change all new tasks to the scheduling technique you wish.

Project allows you to change from Manually Scheduled to Auto Scheduled on a task-by-task basis, so you are not bound to the default. The "Task Mode" column allows you to click and choose whichever method fits your need at the moment.

Figure 2.9: Changing scheduling mode.

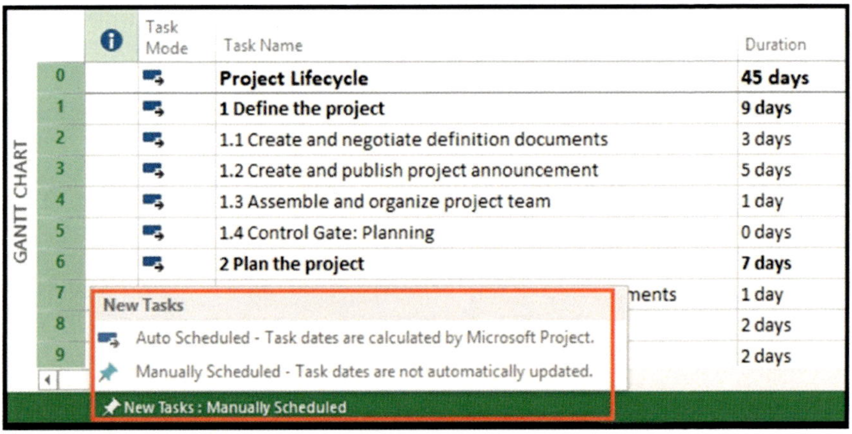

		Task Mode	Task Name	Duration
	0		**Project Lifecycle**	**45 days**
	1		1 Define the project	9 days
	2		1.1 Create and negotiate definition documents	3 days
	3		1.2 Create and publish project announcement	5 days
	4		1.3 Assemble and organize project team	1 day
	5		1.4 Control Gate: Planning	0 days
	6		2 Plan the project	7 days
	7	New Tasks		1 day
	8	Auto Scheduled - Task dates are calculated by Microsoft Project.		2 days
	9	Manually Scheduled - Task dates are not automatically updated.		2 days
		★ New Tasks : Manually Scheduled		

> **Do This:** Stay in Manually Scheduled mode until your initial task list is entered. Consider changing all existing data to Automatic Scheduling just prior to sequencing tasks and assigning resources. Project can then adjust the resource schedule to the task schedule. This will be discussed in more detail later.

Once the project is set up, you are finally ready to prepare the task list and organize it. It takes some thought and planning, but the next step will set the stage for better reporting and scheduling.

Do's and Don'ts Review

Before entering project data, make sure that Microsoft Project is ready to calculate it! Project needs to be given your rules for it to work for you. It does not know how many hours a day you wish to work, nor how many workdays are in your work week. It does not know when the project should start, so it assumes that work begins on the current date. Help it help you by setting your file up for success.

Do's:
- When creating a new project always set the project start date.
- Keep calendars simple at first.
- Schedule from the start date when setting up a project.
- Stay in Manually Scheduled mode until your initial data is entered, then switch to Automatic.

Don'ts:
- Don't start a new project by entering data! Set up and configure Project to produce a more realistic task schedule.
- Don't re-create calendars for every file! Use the Organizer instead. See Appendix A for more on the Organizer.

Chapter 3: Organize the Task List

In this chapter, you will learn:

- Why you should organize your project tasks
- How to design a task outline in Project
- How to estimate and enter task durations

Why You Should Organize Your Task List

Without organization, a project is merely a list of tasks that are required to complete the project. Organization provides structure and is supported by systematic and repeatable practices. It is the key to solid project design and accurate reporting.

In Microsoft Project, organization is imposed by outlining tasks. Any task that has other tasks indented under it is a Summary Task. Indented tasks may also be Sub-Summary Tasks. In other words, the project outline allows for multiple levels of detail. Tasks that are indented under a Summary Task but have no tasks indented under them are Sub-Tasks. For example:

```
Project Summary Task
    Level 1 Summary
        Level 2 Sub Task
        Level 2 Sub Task
    Level 1 Summary
        Level 2 Sub Task
        Level 2 Sub Task
        Level 2 Sub Summary Task
            Level 3 Sub Task
            Level 3 Sub Task
```

Resources will later be assigned to sub-tasks only. Thus, work, costs and tracking are sub-task operations.

Summary tasks are organizational by nature. The total work and cost in all of the related sub-tasks roll up to the summary task where that information is aggregated. The duration of a summary task is not the sum of all sub-task durations. Its duration is the amount of calendar time scheduled for its sub-tasks, from the start of the earliest sub-task to the finish of the latest.

Figure 3.1: Duration of a summary task.

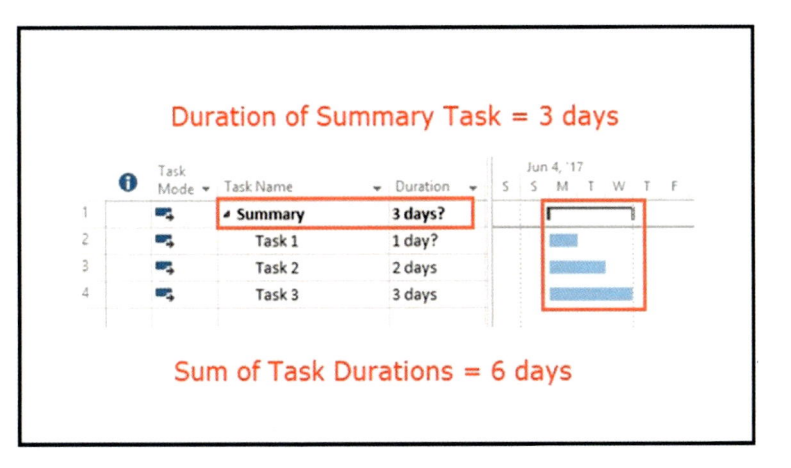

The roll up feature of an outline suggests that a project designed around reporting requirements such as by phase, product or deliverable will automatically aggregate data in the correct "bucket" for reporting purposes.

Figure 3.2: Organizing tasks through outlining.

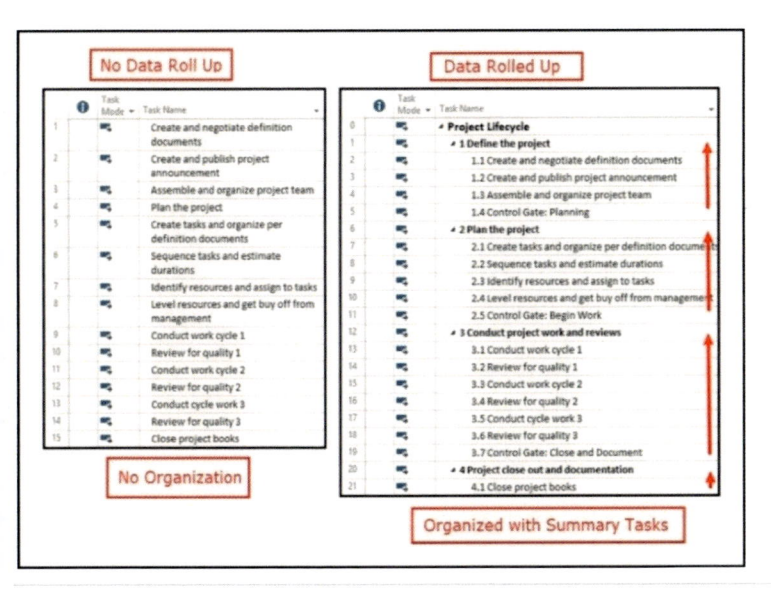

Ch 3: Organize the Task List | 39

To design a project outline, ensure that the outlining tools are turned on.

> ✓ <u>Do This:</u> Turn on your outlining tools before entering any task data! Go to the Format tab and ensure the Project Summary Task and Summary Tasks checkboxes have been checked. If these are turned off you will be unable to create an outline.

Figure 3.3: Turn on outlining tools.

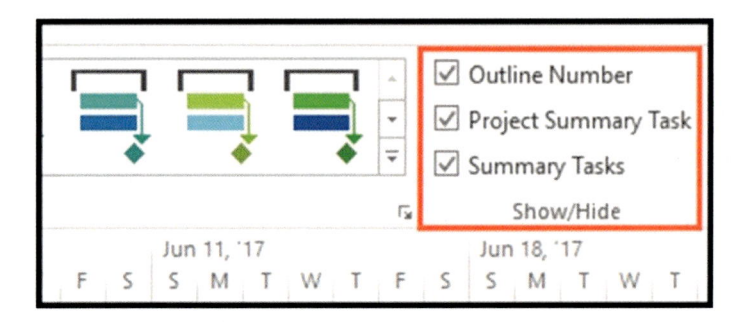

The Project Summary Task will show you the project level details—its Duration, Start date and Finish date. Other project-level information is available in Tables, described earlier.

Figure 3.4: The Project Summary Task receives the roll up of all tasks and summary tasks indented under it.

	Task Name	Duration	Start	Finish
0	▲ **Project Lifecycle**	**46.5 days**	**6/5/2017**	**8/8/2017**
1	▲ **Define the project**	9 days	6/5/2017	6/15/2017
2	Create and negotiate definition documents	3 days	6/5/2017	6/7/2017
3	Create and publish project announcement	5 days	6/8/2017	6/14/2017
4	Assemble and organize project team	1 day	6/15/2017	6/15/2017
5	Control Gate: Planning	0 days	6/15/2017	6/15/2017
6	▲ **Plan the project**	8.5 days	6/16/2017	6/28/2017
7	Create tasks and organize per definition documents	1 day	6/16/2017	6/16/2017
8	Sequence tasks and estimate durations	2.5 days	6/19/2017	6/21/2017
9	Identify resources and assign to tasks	2.5 days	6/21/2017	6/23/2017
10	Level resources and get buy off from management	2.5 days	6/26/2017	6/28/2017
11	Control Gate: Begin Work	0 days	6/28/2017	6/28/2017
12	▲ **Conduct project work and reviews**	24 days	6/28/2017	8/1/2017
13	Conduct work cycle 1	7 days	6/28/2017	7/7/2017
14	Review for quality 1	1 day	7/7/2017	7/10/2017
15	Conduct work cycle 2	7 days	7/10/2017	7/19/2017
16	Review for quality 2	1 day	7/19/2017	7/20/2017
17	Conduct cycle work 3	7 days	7/20/2017	7/31/2017
18	Review for quality 3	1 day	7/31/2017	8/1/2017

Begin with a Top-Down Design

Top down design is another method of organizing project tasks. It is similar to the way we often organize work in real life. Here is an example:

I want to paint the walls in a room in my house. The walls need to be repaired, cleaned, primed and painted. Top-down design allows me to start my outline with the goal—a freshly painted room—as a summary task. The work to achieve that goal is indented and listed under the summary. The result is a top-down outline of the tasks and roll up of work and cost. It would look something like this:

(Summary task)   Paint Room
(Sub task 1)   Wash walls
(Sub task 2)   Repair walls
(Sub task 3)   Prime walls
(Sub task 4)   Paint walls

Think of it as a decomposition of the project from the product or outcome of the project to the individual work items that define it. The top item is the most outdented item—the Project Summary Task. Remember that this feature must be turned on as it is an Outlining feature.

Once the Project Summary Task is in place, you begin inserting the next tier of information, using the buttons from the Insert command group in the Task tab. This tier of data is often mostly summary tasks, though high-level tasks can be placed here as well. This tier often represents the sub-products or major components of the project product. After the summary tasks are defined, the work can be listed as tasks. Using our painting example, the Project Summary Task and Summary tasks might look like this:

Project Summary Task	Paint home interior
(Summary task)	Paint Living Room
(Sub task 1)	Wash walls
(Sub task 2)	Repair walls
(Sub task 3)	Prime walls
(Sub task 4)	Paint walls
(Summary task)	Paint Dining Room
(Sub task 1)	Wash walls
(Sub task 2)	Repair walls
(Sub task 3)	Prime walls
(Sub task 4)	Paint walls

Each room is identified as a summary task and the work to be accomplished is listed as a task. This design is repeated until all tasks have been identified.

Figure 3.5: Insert buttons.

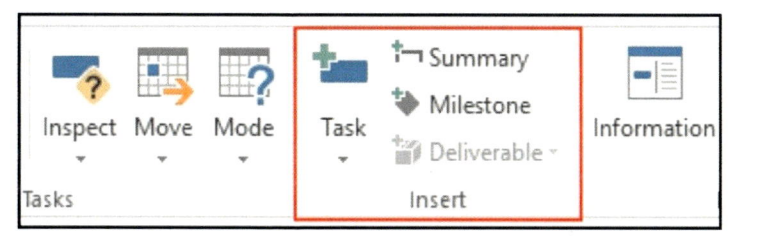

While inserting the tasks, you will need to outdent and indent tasks to create the outline. This is accomplished using the Outdent and Indent buttons found in the Schedule command group of the Task tab.

Figure 3.6: Indent and Outdent buttons.

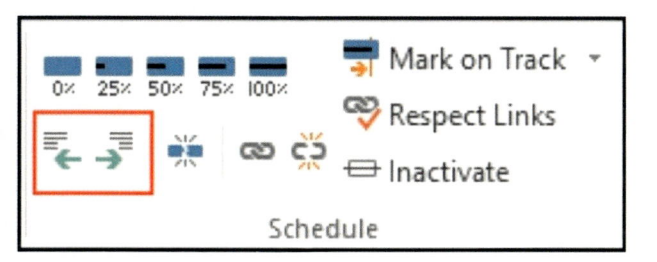

> Do This: Watch the task insertion points carefully! As with other Office applications, Project will always insert a Summary or other task at the same level as the task immediately above it in the task list. Just outdent or indent tasks as needed.

> Do This: Consistency counts! A successful project outline depends on consistency for rolling up data from tasks and summary tasks to your project summary task. Remember that your project outline should follow this general format:

Project Summary Task
    Level 1 Summary
        Level 2 Sub Task
        Level 2 Sub Task
    Level 1 Summary
        Level 2 Sub Task
        Level 2 Sub Task
        Level 2 Sub Summary
            Level 3 Sub Task
            Level 3 Sub Task

Keep indenting and outdenting data until you reach the level of detail in which you can identify individual work, number of resources or calendar time estimated to do the work. In the painting example the walls could have been further broken down to north wall, south wall, east wall and west wall. Instead, the outline stopped at walls. There was no need to be more specific. If the walls were different in composition or color, the decision to list each wall in each room might make sense and provide clarity.

Finish with Bottom-Up Design

This step is a verification that the design is such that the data will roll up through all levels of the project correctly. Look at each summary or sub task and determine that it is at the correct level and organized correctly for reporting purposes. Start at the lowest level in the task list and work your way to the top, correcting as you go by using indenting and outdenting.

> <u>Do This:</u> Organize around reporting requirements! It will simplify reporting, troubleshooting and design while clarifying the structure of the project. Create a summary task for deliverable, phase or any other "bucket" you need to aggregate data for reporting.

Don't Do This: Don't try to identify tasks, durations, dates, cost, resources or work at the same time. At this point in entering data into Project your focus should just be in creating the outline. Resources, work, durations and cost will all be entered or calculated later. Keeping your actions focused and systematic will enable you to design your outline without the other complications.

Do's and Don'ts Review

A properly outlined project in Microsoft Project provides the organization of information being portrayed and a solid structure for reporting.

Do's:
- Turn on your outlining tools before entering any task data.
- Watch the task insertion points carefully.
- Be consistent with your outline. Roll-up requires it.
- Organize around reporting requirements.

Don'ts:
- Don't try to identify tasks, durations, dates, cost, resources or work at the same time. It's confusing.

Chapter 4: Enter Task Durations

In this chapter, you will learn:

- Duration variables
- Schedule variables
- Task types

Once the outline is entered in Microsoft Project, you are ready to estimate and enter task durations. This is accomplished by typing the estimated duration in minutes, hours, days, weeks or months into the Duration field:

- 1 m = 1 minute;
- 1h = 1 hour;
- 1d = 1 day;
- 1w = 1 week; and
- 1mo = 1 month

In Project, duration is a measure of working time, not sequential or calendar time. This means a week is five days. If a task is started on a Friday, it is concluded on the following Thursday since weekends are not working time by default. The default duration is set for days.

Figure 4.1: The Duration column is a field. It indicates the span of working time for a task. On a summary task it is the span of working time from the start of the earliest sub-task to the finish of the latest.

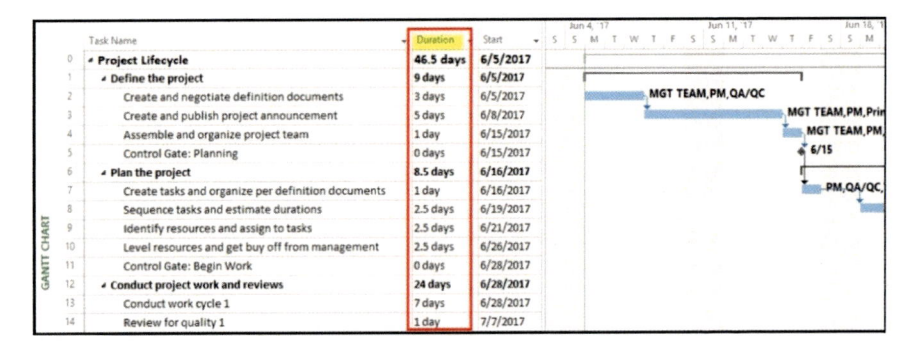

	Task Name	Duration	Start
0	**Project Lifecycle**	**46.5 days**	**6/5/2017**
1	**Define the project**	**9 days**	**6/5/2017**
2	Create and negotiate definition documents	3 days	6/5/2017
3	Create and publish project announcement	5 days	6/8/2017
4	Assemble and organize project team	1 day	6/15/2017
5	Control Gate: Planning	0 days	6/15/2017
6	**Plan the project**	**8.5 days**	**6/16/2017**
7	Create tasks and organize per definition documents	1 day	6/16/2017
8	Sequence tasks and estimate durations	2.5 days	6/19/2017
9	Identify resources and assign to tasks	2.5 days	6/21/2017
10	Level resources and get buy off from management	2.5 days	6/26/2017
11	Control Gate: Begin Work	0 days	6/28/2017
12	**Conduct project work and reviews**	**24 days**	**6/28/2017**
13	Conduct work cycle 1	7 days	6/28/2017
14	Review for quality 1	1 day	7/7/2017

Milestones are zero duration tasks that point out a date in the schedule. No resources are assigned to this type of task in Project. In the schedule, they have the appearance of a diamond, not a bar. One use of a milestone is to denote the end of a phase. Another example is to denote a time in the schedule when a report is due.

Figure 4.2: Milestones have no duration.

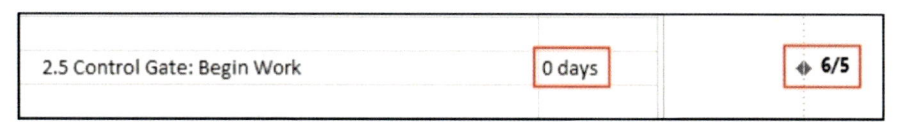

| 2.5 Control Gate: Begin Work | 0 days | ◆ 6/5 |

Before assigning durations to tasks, you need to understand the relationship of all the task schedule variables and how they inter-relate.

✓ **Do This:** At this point, carefully consider changing your scheduling mode to Automatically Schedule. Let Project figure out the finish dates dynamically.

✗ **Don't Do This:** Don't enter a start or finish date for a task in the early stages of creating the project file. Project understands this to be a manual override to calculations and will constrain the task. This hinders dynamic scheduling.

Figure 4.3: Fields used in estimating. Default Duration is in days, Work is expressed in hours and Units of Resource in a percentage. 500% means five people working full time.

	Task Mode	Task Name	Duration	Work	Resource Names	M T W T
0		⁴ Project Lifecycle	46.5 days	1,396 hrs		
1		⁴ Define the project	9 days	176 hrs		
2		Create and negotiate definition documents	3 days	72 hrs	MGT TEAM,PM,QA/QC	MGT
3		Create and publish project announcement	5 days	80 hrs	MGT TEAM,PM,Printer Paper[3 Reams]	
4		Assemble and organize project team	1 day	24 hrs	MGT TEAM,PM,QA/QC	
5		Control Gate: Planning	0 days	0 hrs		
6		⁴ Plan the project	8.5 days	260 hrs		
7		Create tasks and organize per definition documents	1 day	40 hrs	PM,QA/QC,Team[300%]	
8		Sequence tasks and estimate durations	2.5 days	100 hrs	Team[500%]	
9		Identify resources and assign to tasks	2.5 days	80 hrs	MGT TEAM[200%],PM,QA/QC	
10		Level resources and get buy off from management	2.5 days	40 hrs	MGT TEAM,PM	
11		Control Gate: Begin Work	0 days	0 hrs		
12		⁴ Conduct project work and reviews	24 days	864 hrs		
13		Conduct work cycle 1	7 days	280 hrs	Team[500%]	
14		Review for quality 1	1 day	8 hrs	QA/QC	

In Project, these are the fields used in estimating:

Work: The effort in hours.

Units of resource: The amount available and assigned to your project. Usually expressed as a percentage, but can also be thought of in terms of work. In other words, Units can be expressed in different measures. One resource working 8 hours per day full time is 1, 100% or just the work value of 8 hours.

Duration: Calendar time to complete work. Expressed in working days.

Work, units and duration are expressed in simple mathematics wherein:

Work = Units x Duration

In this equation, the unknown is how much work will be associated with a task. To calculate the work, the number of resources and the duration of the task must be known or previously estimated. One person working eight hours per day on a two-day task means that 16 hours of work is expected. Adding a second resource for the two-day period increases the work value to 32 hours.

Units = Work / Duration

The unknown in this case is the number of resources required to complete a task. To calculate the units, work and duration must be known or previously estimated. Duration can be expressed in working days of eight hours each. Work is the total amount of effort expected in the task. If the total effort expected is 24 hours and the duration is one workday of eight hours, then 24 hours of work / eight hours of duration = three units of resource.

Duration = Work / Units of Resource

Duration is the unknown. Work and resource units are known or already estimated. If work is 24 hours and three people work concurrently, the duration will be one day.

Task Types

Each of the above variables can be "Fixed," allowing you to enter a second variable. Project will then take the information provided and calculate the third variable based on the Task type. Task types are found in the Task Information dialog and in the Task form. Task types are only available for automatically scheduled tasks.

> ✅ <u>Do This:</u> Use Help to review and refine the Task types required in your project! Help contains a full discussion on Task types and how Project reacts to changes in work, duration and units. See Appendix B for more details.

Figure 4.4: Help is brought up by pressing F1 on the keyboard or by typing in the "Tell me what you want to do" box. See Appendix B.

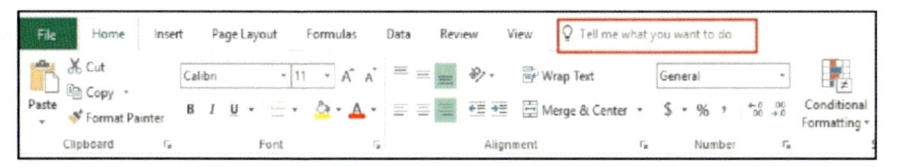

Figure 4.5: Task types found in the Task Information dialog.

The default task type is Fixed Units. Work is the result of the units multiplied by the duration entered. Project will not adjust the units of resource in a Fixed Units task. If the "Effort-driven" box is checked, the duration will be adjusted by Project based on the work assigned to the task.

Figure 4.6: Task types found in the Task form.

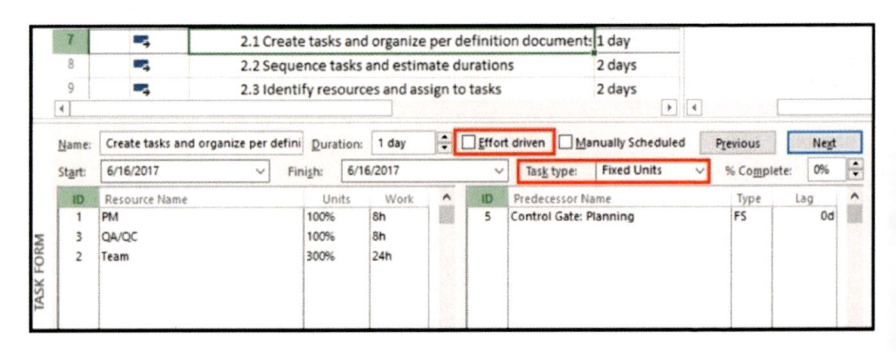

There are no hard rules for choosing a task type. There are, however, "rules of thumb" that are suggested in Project Help:

- Use Fixed Units when you know how many resources will be assigned the task and will remain assigned to the task even if the duration changes because the amount of work changes.
- Use Fixed Work if you want to adjust task duration by assigning or removing resources.
- Use Fixed Duration when a task's duration is known or expected but resource assignments and the amount of work is unknown.

Here are a few examples of applying task types:

Fixed Duration: You have five days to get the task work done, and the amount of work doesn't control the duration. In this instance, you add as many people as required to accomplish the task. A fixed duration task will not increase duration even if the Effort-driven box is checked.

Fixed Units: Your task is estimated at five days. Assuming an eight-hour workday, one person is assigned, resulting in 40 hours of work. If the Effort-driven box is left unchecked, adding a second resource will add another 40 hours for a task total of 80 hours, but the duration of the task will remain five days. If checked, the duration will be reduced to 2.5 days because task work will not increase as resources are added to an effort-driven task.

Fixed Work: Your organization uses work to drive schedules, not duration. In this task type, you enter the work estimate and assign resources to the task. The duration is then calculated in a fixed-units, effort-driven task.

Once the durations are entered, you can sequence tasks and get the first look at your task schedule.

Estimating the duration of tasks is one of the early steps in scheduling. Once sequencing of tasks is imposed, the duration and sequence of all project tasks will result in the "network diagram." This view illustrates the schedule flexibility in your project. Inflexible tasks are depicted as the critical path and are formatted in red.

Figure 4.7: The Network Diagram. Red formatting means a task has an inflexible start date.

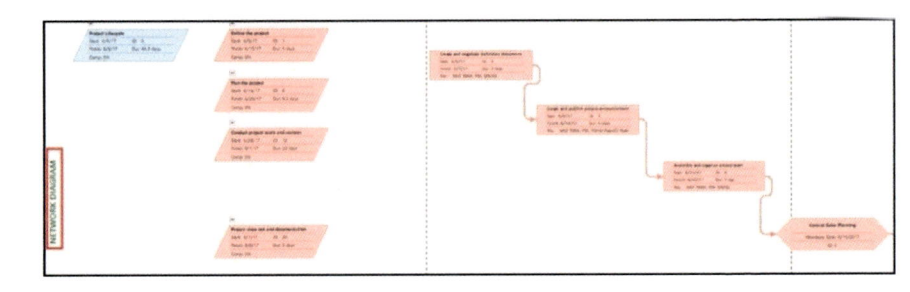

Do's and Don'ts Review

Do's:
- Once the outline is entered, consider changing your scheduling mode to "Automatically Schedule."
- Use Help to review and refine the Task types required in your project!

Don'ts:
- Don't enter a start or finish date for a task in the early stages of creating the project file.

Chapter 5: Sequence Tasks in the Task List

In this chapter, you will learn:

- Why to sequence tasks
- Types of task sequences
- Lead and lag
- How to enter dependencies
- The use of Task Path

Sequencing tasks is a serious step in scheduling. To fully represent the relationships between tasks, remember that sequencing is most effective when real task relationships are imposed without resource constraints. Occasionally, task relationships are a matter of convenience in schedule timing. For example, if you need for tasks to run concurrently and resources will have no schedule conflicts, the tasks can be linked to start or finish at the same time. Later we will see the impact of resource availability to this first schedule when the resource team is assigned to project tasks.

It pays to know what tools Microsoft Project has to assist you in the on-time delivery of your project. We will start by going over several critical definitions.

Predecessor: A predecessor task drives the start or finish of another task. The predecessor does not necessarily precede the task it is driving in the timeline. The two tasks could be concurrent!

Successor: A successor task has its start or finish date driven by another task. It does not necessarily follow its predecessor in the timeline.

Sequencing: This is the term used to describe the process of setting a predecessor or successor relationship between tasks. It's also known as "linking."

Figure 5.1: Types of sequence.

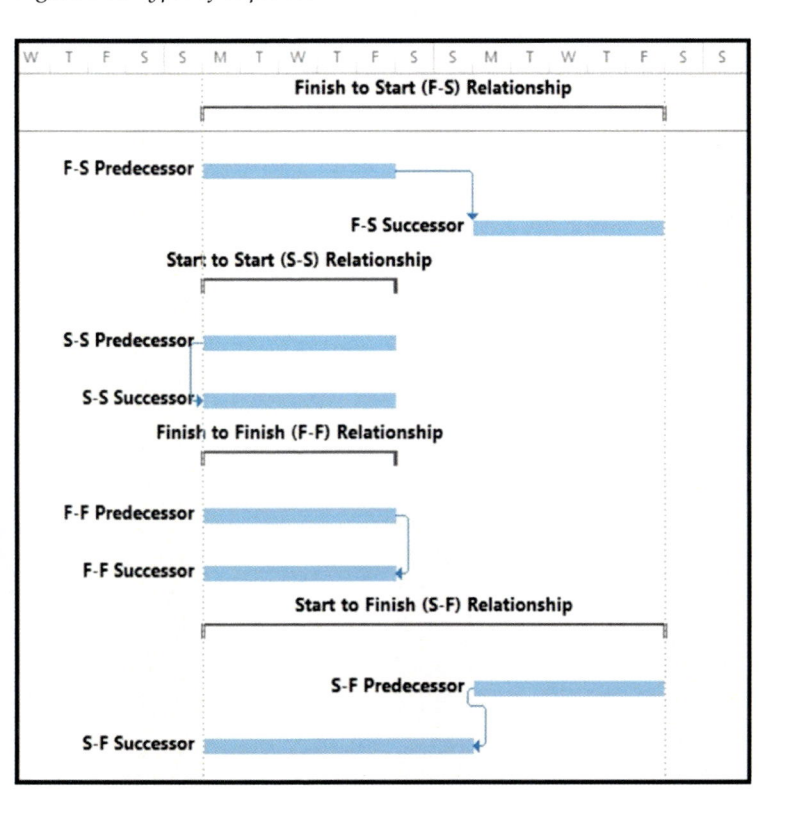

There are four types of sequencing available in Project:

Finish to Start (F-S): The finish date of the predecessor drives the start of the successor. This is Project's default. It represents the most common type of task dependency. The room painting example used in Chapter 3 could require several if not all of the sequence types. Washing the walls in the dining room and letting them dry is certainly the predecessor to what succeeds those activities: painting the walls.

Start to Start (S-S): The start date of the predecessor drives the start of the successor. This is used when tasks are related but can be concurrent without resource conflict. If more than one person is working on refreshing the dining and living rooms, they could each work in a room concurrently washing walls or painting. In this case it is a schedule convenience to establish the start-to-start relationship. Either task could be the predecessor. The concurrency is the goal of the relationship.

Finish to Finish (F-F): The finish date of the predecessor drives the finish of the successor. This is often used for parallel tasks resulting in a handoff or delivery. For example, the painting in both rooms should be complete before brush clean-up begins. The finish-to-finish relationship ensures both paint tasks are completed before clean-up begins.

Start to Finish (S-F): The start date of the predecessor drives the finish date of a successor. This is used in event planning when a task in the project is driving the schedule and when tasks in the sequence need to move as a block rather than individually. As an example of this, final paint touchup could be conducted right up to the point that the room furniture is delivered. Once the delivery starts (the successor), painting (the predecessor) ceases.

Sequencing can be modified to delay the task schedule intentionally. This is known as "lag." Changes to the sequence could also result in overlapping tasks. This is known as "lead" or "negative lag." In the painting example lag might be used to address the time delay required for the walls to dry fully before applying the paint. If an allowance of two hours is given to allow the walls to dry after washing, the relationship between washing the walls and painting would be:

Finish to Start plus two hours lag between tasks

In other words, the successor must wait two hours before starting the task of painting. If the walls dry quickly, the tasks might overlap instead of allowing for delay. This overlap is the opposite of lag, often referred to as "lead."

Do This: Always adjust or refine dependencies in the *successor* task. In sequencing the successor task is dynamically driven by relationships with a predecessor. Lead, lag and the type of relationship aid the successor task to represent the sequence and ultimately the task schedule.

You adjust lead and lag in the Predecessors tab in the Task Information dialog of the Successor Task. Lag is a positive number while lead is a negative number. Lead and lag can be represented as a number of days or as a percentage.

Figure 5.2: Lag shown in the Task Information dialog.

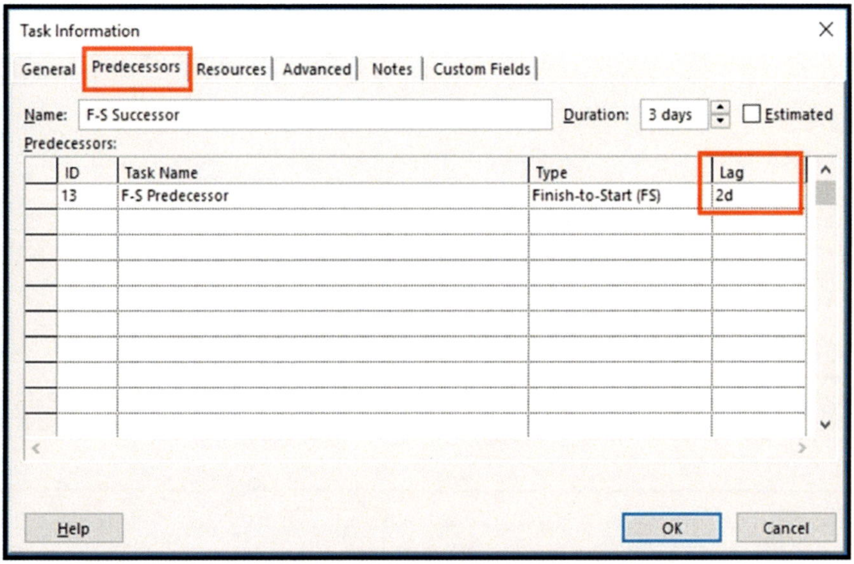

Figure 5.3: Lag shown in the Gantt chart and Predecessors field.

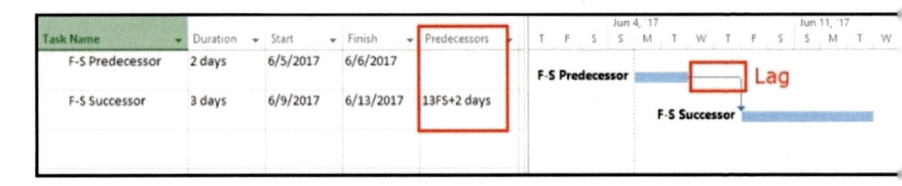

Figure 5.4: Lead shown in the Task Information dialog.

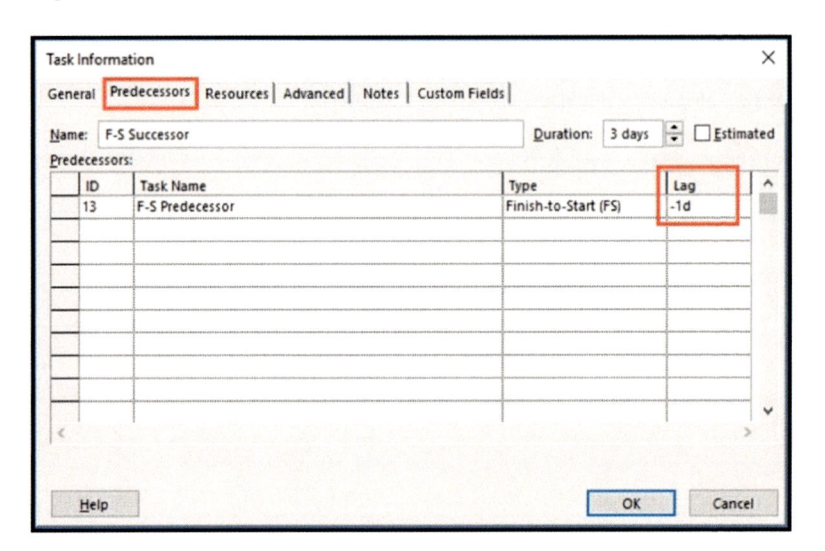

Figure 5.5: Lead shown in the Gantt chart and Predecessors field.

Entering Dependencies

The Predecessors field in the Gantt chart is the most accessible location for entering the default finish-to-start dependency. If a task has a predecessor, you simply type the ID number of the predecessor in the field, press the Enter key and the link is created. Using this method can be quick and comprehensive. Project will assume that the number typed into the Predecessors field is a finish-to-start sequence by default. If you wish to modify the sequence or use a different type, you have to explicitly enter the predecessor's ID number and then the type of sequence such as FS, FF, SS, or SF. If lag is needed, a "+" is entered after the type along with the amount of lag required. For example, "13FS+2days" means task 13 is the predecessor, the type is Finish-to-Start, plus two days' delay between the finish of the predecessor and the start of the successor. If two days' lead is needed, the notation would be "13FS-2days."

Figure 5.6: The Predecessors field describes the exact relationship between tasks, including lead and lag.

	Task Name	Duration	Start	Finish	Predecessors
13	F-S Predecessor	2 days	6/5/2017	6/6/2017	
14	F-S Successor	3 days	6/9/2017	6/13/2017	13FS+2 days

Another common method of linking is to select the tasks you wish to link and click on the Link button in the Schedule command group of the Task tab. Project will link the tasks in the default finish to start relationship.

Figure 5.7: Link and unlink buttons.

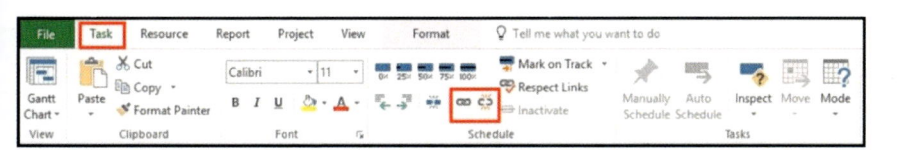

❌ Don't Do This: Don't link summary tasks! Doing so can artificially extend your project duration.

A common characteristic of project schedules is the use of multiple predecessors in task sequencing. The general rule of thumb is that the latest predecessor's dates drive the successors. Finding which predecessor that is can be a time-consuming bit of analysis.

✅ Do This: Use "Task Path" to clarify the predecessors driving and successor driven in relationships. It will save you time and frustration.

Starting with Project 2013, Microsoft has made this much easier with a new feature: Task Path. Using Task Path, you can select any task in the task list and find out which tasks are driving its schedule dates. Found in the Format tab, Task Path uses a different bar color format to identify "Driving Predecessors."

Figure 5.8: Task Path aids in analyzing task relationships.

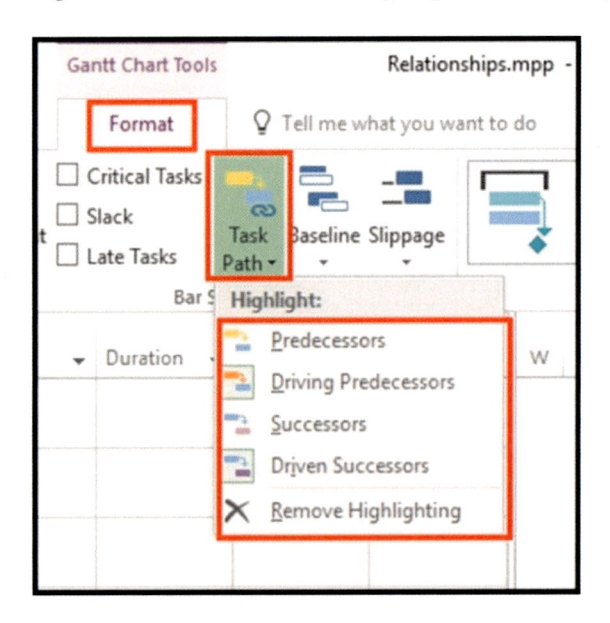

Task Path will identify the predecessor that is latest in the schedule and identify it as the "Driving Predecessor" by changing the task bar color to orange.

Figure 5.9: Task Path with Driving Predecessors turned on.

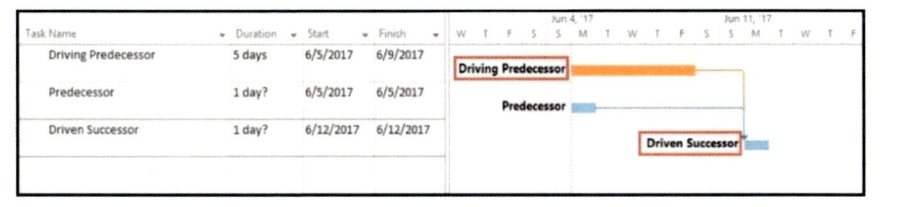

Task Name	Duration	Start	Finish
Driving Predecessor	5 days	6/5/2017	6/9/2017
Predecessor	1 day?	6/5/2017	6/5/2017
Driven Successor	1 day?	6/12/2017	6/12/2017

Task Path will identify the successor task being driven with a purple color.

Figure 5.10: Task Path with Driven Successors turned on.

Task Name	Duration	Start	Finish
Driving Predecessor	5 days	6/5/2017	6/9/2017
Predecessor	1 day?	6/5/2017	6/5/2017
Driven Successor	1 day?	6/12/2017	6/12/2017

Do's and Don'ts Review

Sequencing tasks is essential in scheduling work to occur at the right time in a project. Along with duration, sequencing represents your preferences in achieving schedule goals.

Do's:
- Always adjust or refine dependencies in the successor task.
- Use "Task Path" to clarify the predecessors driving and successor driven in relationships.

Don'ts:
- Don't link summary tasks.

Chapter 6: Create Resources and Assign them to Tasks

In this chapter, you will learn:

- Resource types
- How to create resources
- How to assign the resource types
- How to evaluate resource schedules
- How to level resources
- Methods of leveling

If your project is small or has few resources, you may wish to avoid the complexities that "resources" bring to the project plan. Many Microsoft Project users have found that a total work threshold assists in determining when resourcing a project is necessary. For example, perhaps the company does not wish to get reports or track projects that contain fewer than 100 hours of effort. In this manner, only projects over 100 hours of total effort will require management rigor and oversight.

It requires time, effort and patience to manage the resourcing of a Project file. Is it wise to spend 20 hours of effort managing 20 hours of project work?

> Do This: Consider the project's total work estimate prior to committing resources. Is there enough work to justify the overhead of managing resources in the Project file?

If you must share resources, another of Project's capabilities should be considered: the use of a "resource pool." A resource pool is a Project file that shares its resources with other Project files. When you use a resource pool, you can view resource assignments across all Project files sharing the resources. This means you can also analyze and modify the resources work and cost across all of the Project files sharing from the pool. This can be a handy model in organizations that require a simple project management process.

"Subprojects" are another capability of Project. This is the embedding of multiple projects into a single Project file and consolidating the resources into a single pool. Sometimes referred to as a "master project," this model offers the benefits of a resource pool with the entire task list of all embedded projects in one file. It also offers cross-project sequencing and reporting. In construction, many projects are composed of smaller projects. For example, home builders seldom have the personnel and skills available on staff to complete a home construction project. They must rely on sub-contractors to do the specialty work. A beautiful new home is often the result of many small projects implemented at the appropriate time in the building process. The home is the product of the sub-projects.

Resource pools and master projects are beyond the scope of this book, but you can find details in Project's Help and get help in the MPUG Discussion forums or in the Microsoft Project Forums within Microsoft TechCenter. Both are provided at no cost. The link to the MPUG Discussion forum is www.mpug.com/forums/type/discussion/.

The link to the Project TechCenter is https://social.technet.microsoft.com/Forums/projectserver/en-US/home.

Regardless of the different models available to you, the decision to use resources requires knowledge about the feature and its use. Let's go over the types of resources used by Project.

Resource Types

There are three principal types of resources used in Project: working, material and cost.

Work resource: Performs work and can have a cost rate. A person or rental equipment are examples of this type resource. This is also the default type of resource in Project.

The cost of a work resource is determined by the amount of work multiplied by its rate, usually expressed in hours: Pat is a plumber and earns $75 per hour. If Pat works on a task for two hours, Pat's labor cost is $150.

Material resource: This type of resource can be identified with a unit cost. Paint, gasoline, lumber and concrete are examples of material resources. For example, Pat the plumber needs 20 lineal feet of tubing to conduct a task. The tubing is $2 per lineal foot, so the cost of the tubing is $40.

Cost resource: This type of resource does no work and has no unit cost but does have a cost in the project. Permits, lodging and airfare are examples of cost resources. As an example, for Pat to begin the plumbing repairs, a permit is required. The municipality charges a $75 minimum for this type of repair. The permit fee will be assigned to Pat's task at $75.

Let's summarize those costs. Pat's task requires two hours of effort ($150), a permit ($75) and 20 lineal feet of tubing ($40). The total cost of the task is $265.

How to Create Resources in Project

In the View tab in the Resource Views command group, bring up the Resource sheet and select the first empty cell in the Resource Name field to begin defining resources.

Figure 6.1: The Resource sheet.

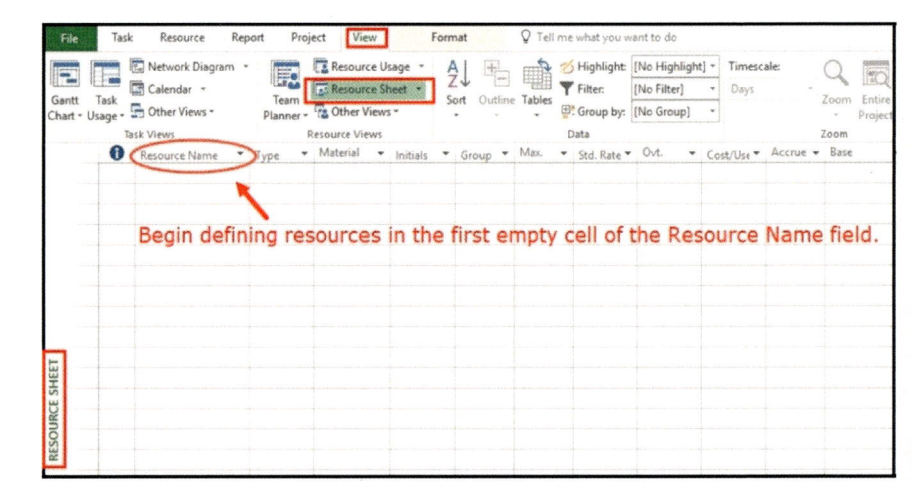

Switch tabs to the Resource tab. This tab contains the tools required to define resources. Select the Add Resources Button from the Insert command group. Project will offer you a list of available resource types.

Don't Do This: Don't define resources in views other than the Resource sheet. If you do, the chance of resources being duplicated somewhere else in your project plan is high.

Figure 6.2: Use the "Add Resources" button to build your project team.

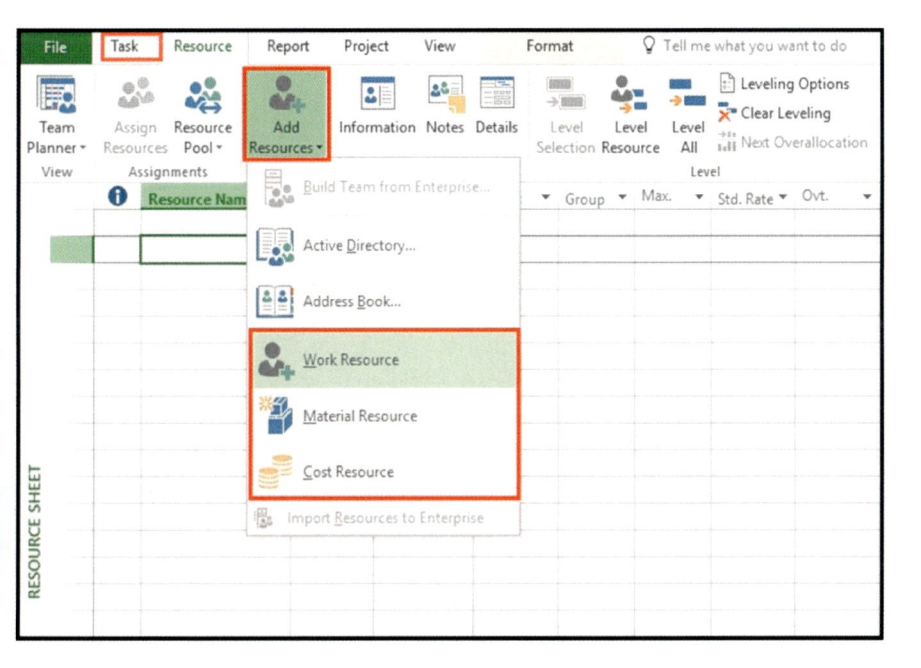

Click on the resource type you wish to create. Project will place the new resource in the Resource sheet.

Figure 6.3: The newly inserted resource.

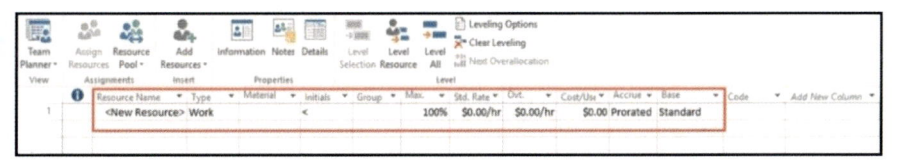

Next, at a minimum, define the Resource Name, Initials, Max Units (how much of a resource is available in units of 1 or 100 percent initially, depending on your settings), Group and Rate.

Ch 6: create Resources & Assign Tasks | 71

If the resources are Material resources, enter a title representing the basis for the unit cost into the "Material Label" field. For example, the basis of the cost resource might be "per gallon (s)" if it is paint or "per ream(s)" if it is paper.

If the resource is a cost resource, the cost is not entered into the Resource sheet. The cost will be entered each time the resource is assigned a task.

Figure 6.4: The Resource sheet showing resource types.

Resource Name	Type	Material Label	Initials	Group	Max.	Std. Rate	Ovt.	Cost/Use	Accrue	Base
Work Resource	Work		W		100%	$100.00/hr	$0.00/hr	$0.00	Prorated	Standard
Material Resource	Material	unit cost each	M			$10.00		$0.00	Prorated	
Cost Resource	Cost		C						Prorated	

> ✅ Do This: Always adjust resource Max Units to the amount of resource that will be available at any point in the project schedule. If the resource is a group, the Max Units should reflect the group's total membership. For example, entering "5" or "500%" in Max Units represents a group of five members. Project will not keep track of any specific member of the five, just the total membership. If you need to track each resource individually, then each member of the group would be a separate entry in the Resource sheet.

How to Assign Resource Types

> ❌ Don't Do This: Resources should not be assigned to summary tasks. When assigned to a summary task, the resource work is spread through the duration of the task. If the resource is also assigned to a task within the summary, it will double up the work for that resource, creating a problem that will not be easily identified.

Work, cost and material resources are all assigned to tasks from the "Assign Resources" dialog. Each type is assigned in a different manner in the dialog and controlled using the dialog's features.

In your Gantt chart view this dialog is found in the Resource tab in the Assignments command group.

Figure 6.5: The Assign Resources dialog.

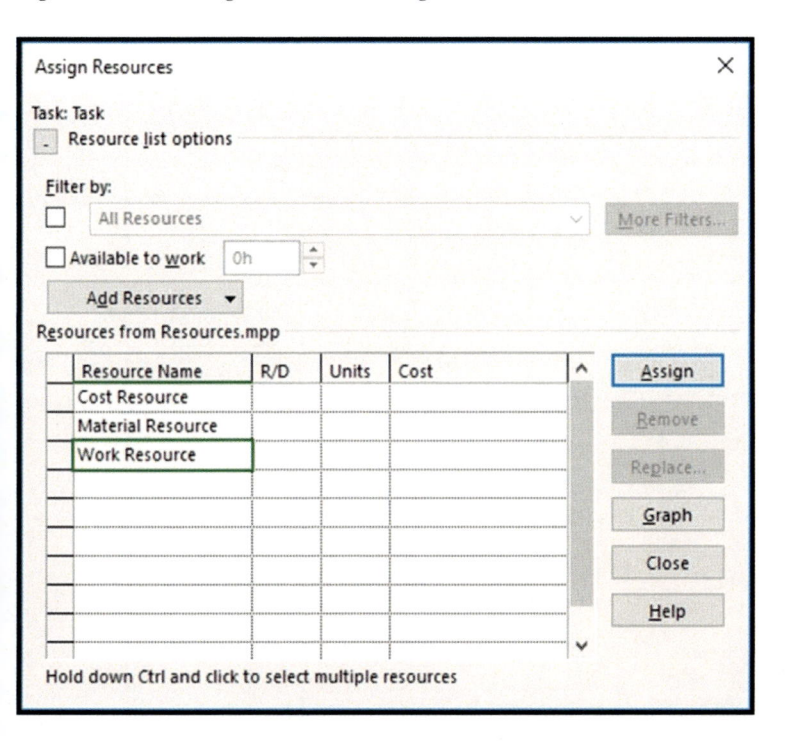

You can work in your Gantt chart while using this dialog. The dialog will stay on top of the Gantt chart while you select tasks, check Task Information or perform task-related editing.

The Assign Resources dialog has the capability of filtering the resource list by type, allowing you to find specific resources quickly. If your resource list is long, this can be a time saver.

Figure 6.6: The Assign Resources dialog showing filters.

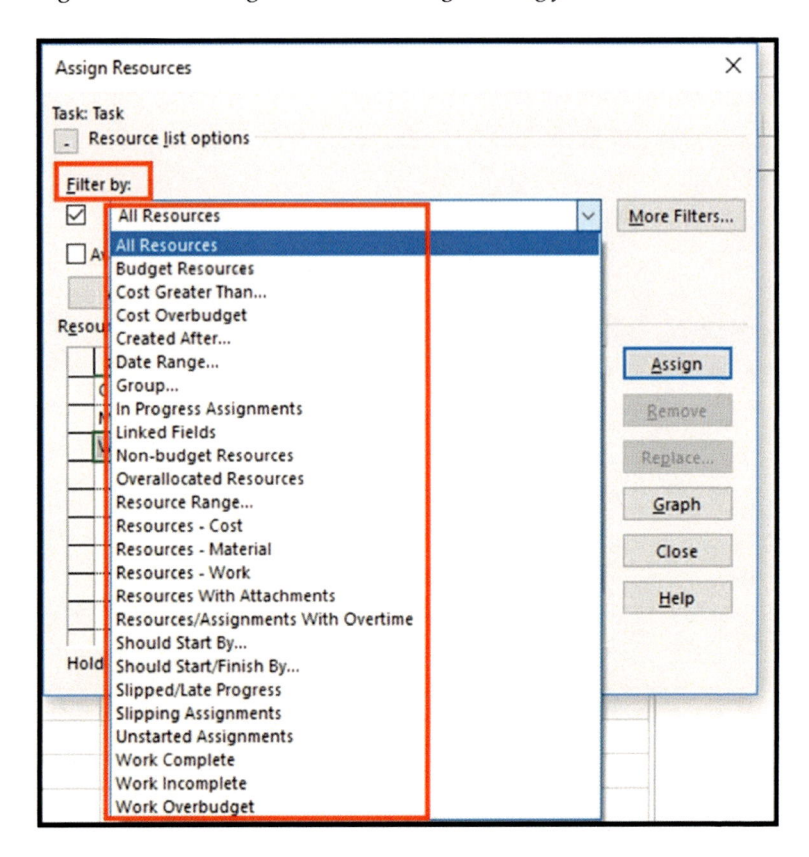

When assigning work resources, select the task from the task list, then select the resource from the resources shown in the form's resource list. If you need to use units that are less than 100%, type the percentage required in the Units column for the resource and click on the Assign button.

If you are assigning a material resource, select the task from the task list, then select the resource from the resources shown in the form's resource list. Type the number required in the Units column for the resource. Then click on the Assign button.

If assigning a cost resource, select the task from the task list, then select the resource from the resources shown in the form's resource list. Type the cost of the resource into the Cost field.

Figure 6.7: The Assign Resources dialog with each type indicated and where the form will need user entry.

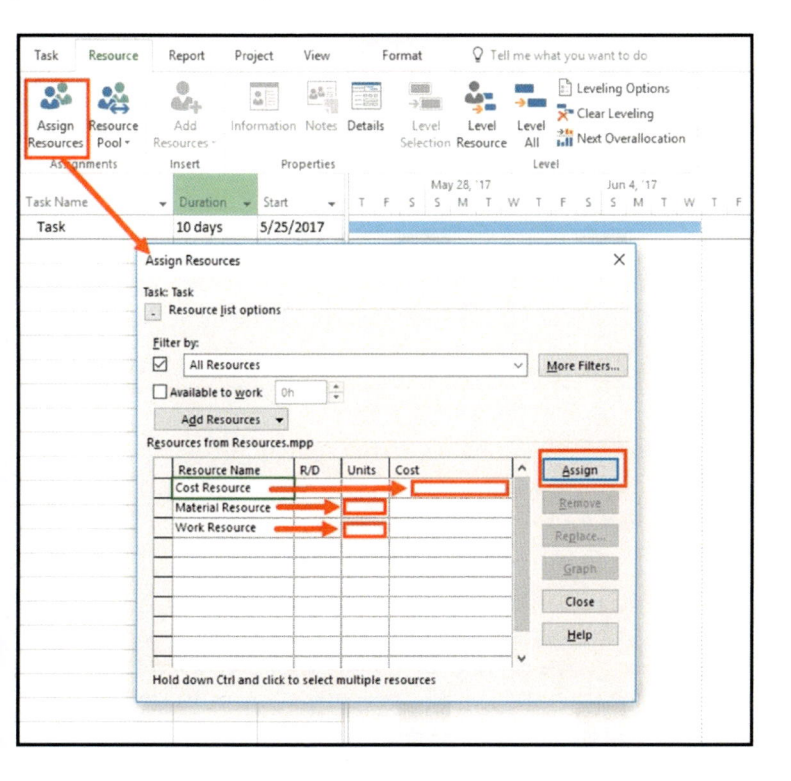

If multiple tasks or resources need to be selected, depress and hold down the Ctrl key while selecting the tasks or resources required in the assignment.

After the assignment is completed, the resources will be listed to the right of the task bar in the Gantt chart.

Figure 6.8: The assignment of a work, material and cost resource showing the different measures of the assignment.

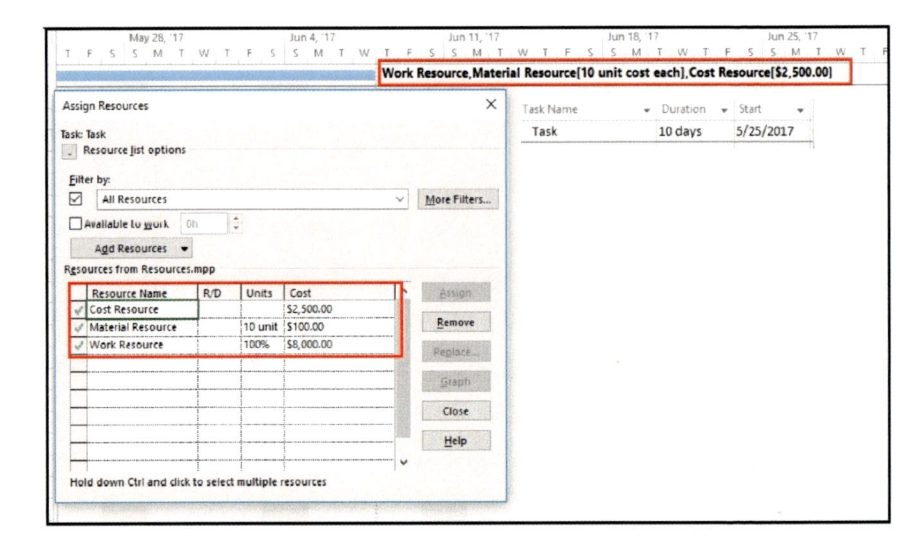

> **Do This:** Check the information to the right of the task bar frequently during resource assignment. Better to find assignment errors early before they become serious.

Evaluating Resource Schedules

Once the resource assignments are made, many project managers conduct an evaluation to ensure that the work is scheduled correctly. This is an important step for avoiding serious problems early in a project's lifecycle. There are several view combinations that can help you evaluate each resource's work load. They are not hard to create, and the information they provide can be priceless.

The Gantt chart is the first view to examine. On the left side of the task sheet an indicator will identify any task that is overallocated. The indicator is a red figure in the Indicators column.

Figure 6.9: The indicator for overallocated resources.

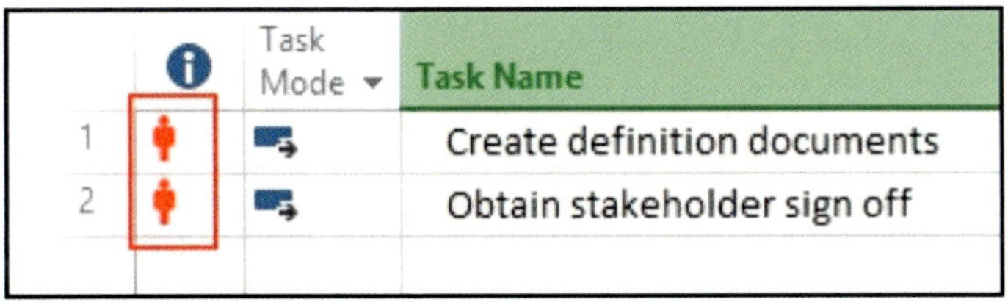

	Task Mode ▼	Task Name
1	🠖	Create definition documents
2	🠖	Obtain stakeholder sign off

This indicator makes it easy to find out where overallocations are occurring in the list of tasks. The second view to examine is the Resource sheet. To evaluate this, you will have to add a column to get the information needed. The Max Units column in the Resource sheet will tell you how many of each resource is assigned to the project, but not how many are required. That data is held in the Peak column. Once the column is inserted and next to the Max Units column, you can compare the number of resources available against the number required in the current schedule. If the numbers in each column are the same, there are no scheduling errors. If Peak is higher than Max Units, the resource is said to be overallocated. This means more work is assigned to the resource than can be scheduled for them based on their calendar settings.

Figure 6.10: A Resource sheet comparing Max Units to Peak for an overallocated resource.

	Resource Name	Type ▼	Material Label ▼	Initials ▼	Group
1	Work Resource	Work		W	
2	Material Resource	Material	unit cost each	M	
3	Cost Resource	Cost		C	

Max. Units ▼	Peak ▼	Std. Rate ▼	Ovt. ▼	Cost/Use ▼	Accrue ▼	Base
100%	200%	$100.00/hr	$0.00/hr	$0.00	Prorated	Standard
		each/day	$10.00	$0.00	Prorated	
		0%			Prorated	

Ch 6: create Resources & Assign Tasks | 77

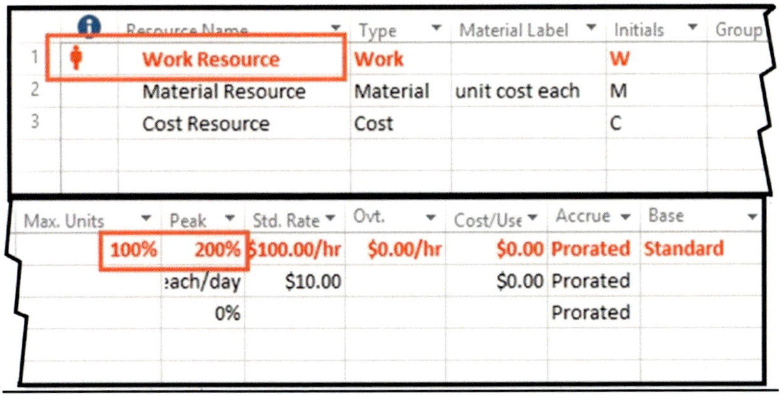

> **Do This:** Insert the Peak column next to the Max Units column. Select "Add New Column" from the right side of the Resource sheet and choose Peak from the list. This will show the Peak column in the sheet. Next, click and hold the Peak column name until you see a four-headed mouse cursor. This indicates that you can drag the column to the position you need in the sheet. Drag the column next to the Max Units column and compare the numbers.

If the numbers match or Peak is lower than Max Units, your resource is within their calendar limitations. If the reverse is true and Peak is higher than Max Units, the resource will need to be leveled. Leveling is a technique to resolve resource over-scheduling, and means taking action to bring the Peak down to Max Units. Another visual hint provided by Project is in the sheet's formatting. Any resource in the list showing a red format is overallocated and needs to be "leveled."

Resource Leveling

When resources are overallocated, the resource schedule is no longer tenable. In most cases overallocation occurs when a resource is assigned to several concurrent tasks.

Figure 6.11: Resource Graph over Gantt showing concurrent tasks causing an overallocation.

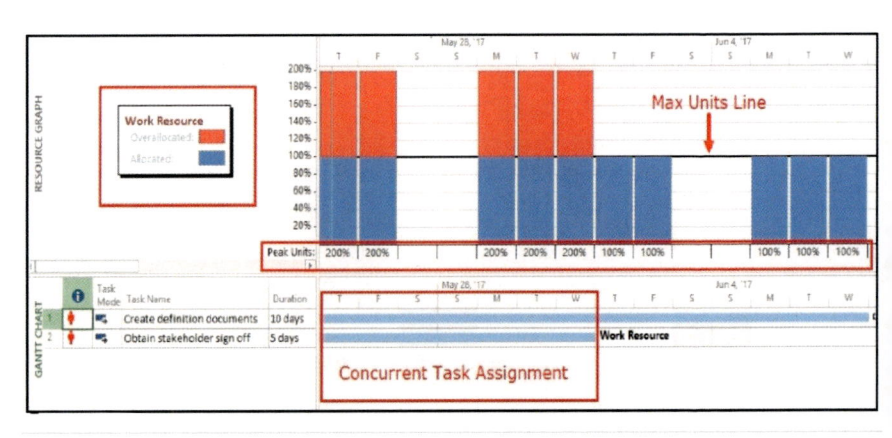

To correct the situation, project managers must employ tactics to reduce the resource workload to a level that can be accomplished. Commonly-used tactics for "leveling" include:

- Delaying tasks until the resource is available.
- Splitting tasks over periods of non-availability.
- Swapping overallocated resources with others that have the same or similar skills and who are available for the duration of the task.
- Deleting tasks.
- Splitting the resource assignment such that the overallocated resource is swapped out at the point of overallocation within the task duration.

Although these tactics often change the cost, schedule or work in some way, leveling can help revise an unworkable schedule to one that has the possibility of success.

Methods of Leveling

The best technique for ensuring that a project is leveled is to not make the mistake to begin with. Well thought out, planned and careful assignments offer the best chance at a realistic schedule. Unfortunately, when a project is large, long, complicated and overallocated even before a project manager is assigned, the damage is already done. Then the project manager needs help in resolving problems quickly. That is also when Project's leveling engine can assist in correcting the problems that are creating the overallocations.

Options Used by Project's Leveling Algorithm

You will find the options by clicking on the Resource tab and choosing Leveling Options in the Level command group. This action brings up the Resource Leveling dialog.

The Resource Leveling dialog is divided into three distinct areas:

- Leveling calculations;
- Leveling range; and
- Resolving overallocations.

Figure 6.12: The Leveling Options dialog identifying the three leveling divisions.

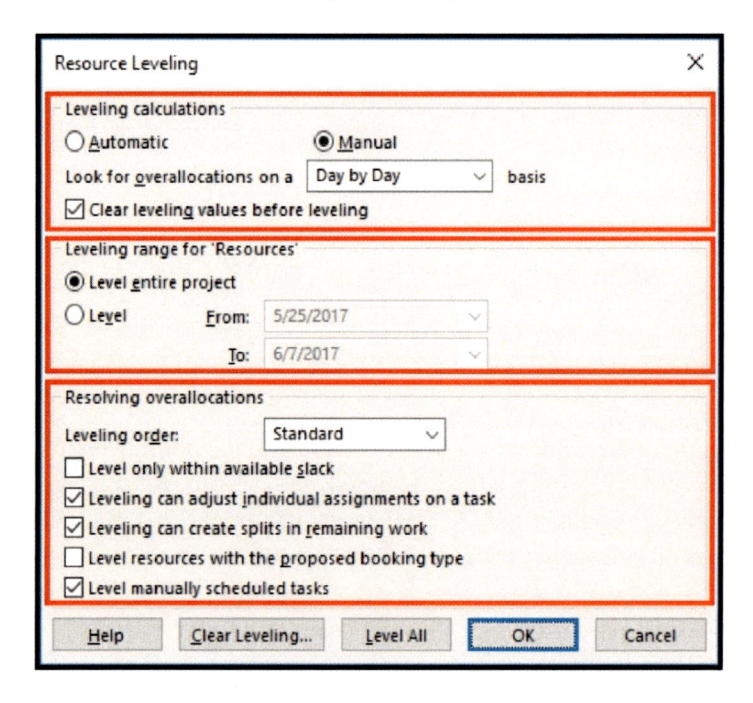

Leveling Calculations Settings

The first selections here are Automatic or Manual. Automatic will prevent overallocations from happening, but may not give you the optimum schedule. Project will level every time you press the Enter key. Manual means that you select the point in time to level by clicking on one of Project's leveling buttons. It does not mean that complicated analysis and user actions produce a leveling solution. Manual is the default.

✓ **Do This:** Leave your leveling calculations set to Manual! When leveling is set to Automatic, depending on other settings you have in place, Project may delay tasks you do not want delayed.

The next setting addresses the granularity of overallocation that Project will look for. Very short or intense projects may need to be leveled on a minute basis, while long projects may just require a monthly adjustment. The default is a day-by-day basis. If any task has an overallocation on any day within the project duration, Project is allowed to level the project, task or resource. Day by day works well for most users.

Figure 6.13: The Resource Leveling dialog with the "Look for overallocations" list shown.

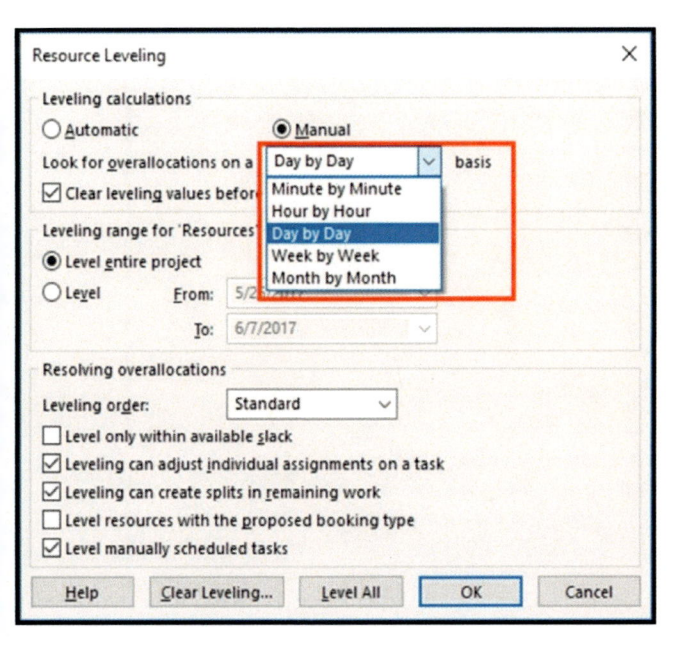

The final setting in Leveling calculations concerns whether previous leveling attempts are to be allowed or not. If delays were required to level in previous attempts, you may wish to leave those delays in place and to add to those delays to accommodate resource schedule conflicts. If that is the case, clear the checkbox, "Clear leveling values before leveling." Then, when a fresh attempt at leveling is needed, Project will eliminate all delays in the schedule and level the project again. This is the default and works well for most users.

Figure 6.14: The Resource Leveling dialog with the "Clear leveling values" checkbox indicated.

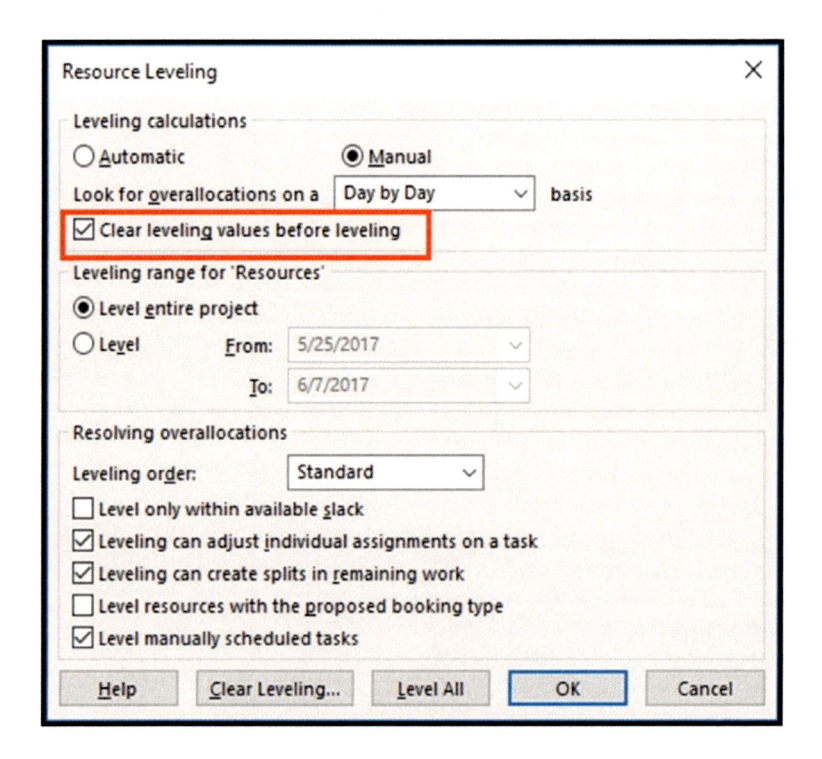

Also helpful are combination views showing the resource graph over a Gantt or "Leveling Gantt," as I mentioned earlier. The Leveling Gantt shows the position of Gantt bars before and after leveling. It also illustrates how much delay Project had to impose on a task to achieve a leveled schedule.

Figure 6.15: The Combination View Resource graph over the Leveling Gantt. Leveling Delay is indicated in the Leveling Gantt.

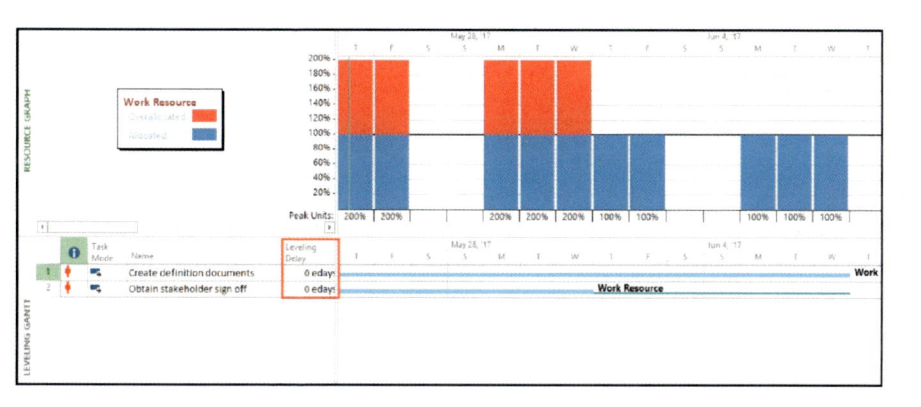

Leveling Range Settings

The leveling range settings are used when a narrow range of dates in the project contain overallocated tasks. In that situation, you may want to level only within that date range. You may want to increase the date range when you use this option to give Project scheduling room to solve the problem. If the entire project is overallocated, you probably need to level all tasks over the duration of the project.

Figure 6.16: The Resource Leveling dialog indicating the leveling range dates.

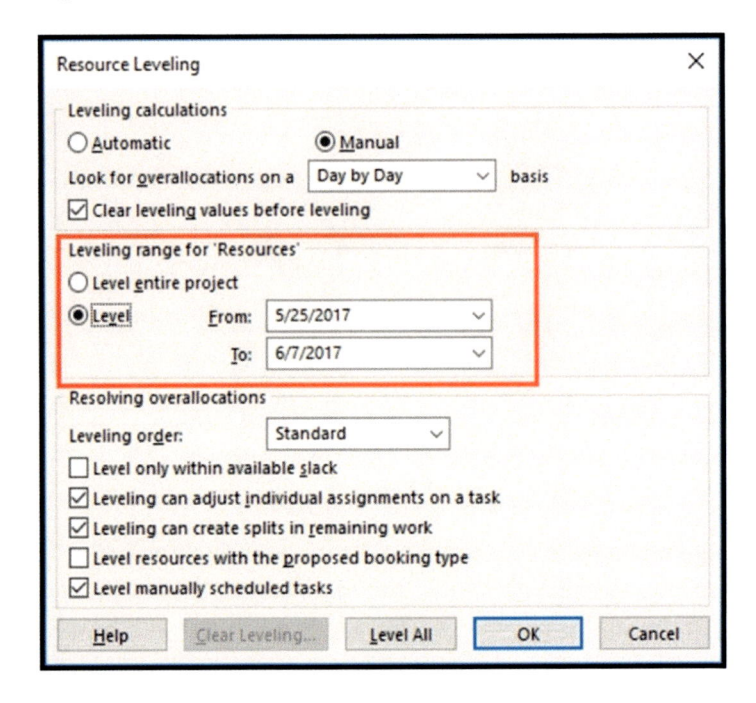

Resolving Overallocations Settings

Leveling order follows three priorities: "ID Only," "Standard" and "Priority, Standard."

Figure 6.17: The Resource Leveling dialog illustrating Leveling order selections.

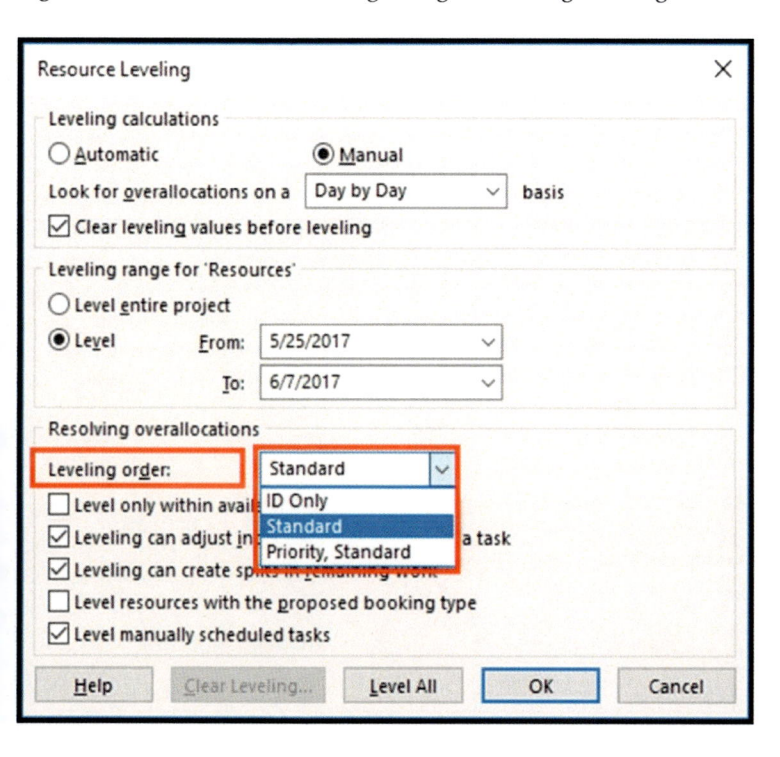

ID Only is the simplest of the three. In the case of overallocation, Project delays tasks that are a higher ID number. For example, if tasks 1 and 2 are scheduled concurrently with the same resource assigned, then task 2 would be delayed to resolve the overallocation.

Standard considers relationships, slack, constraints and priorities before delaying any task having the resource conflict.

Priority, Standard considers the task priority first, then relationships, slack and constraints before delaying any task having resource conflicts. Priority is a task setting chosen in any task's Task Information dialog. It is a value between 1 and 1000. The higher the number, the less likely Project will delay the task. The default priority assigned to every task is 500.

The Task Information dialog box is shown when a task in the task list is double-clicked. It is available from the General tab and is directly under the Duration field.

Figure 6.18: The Task Information dialog indicating task Priority in the General tab.

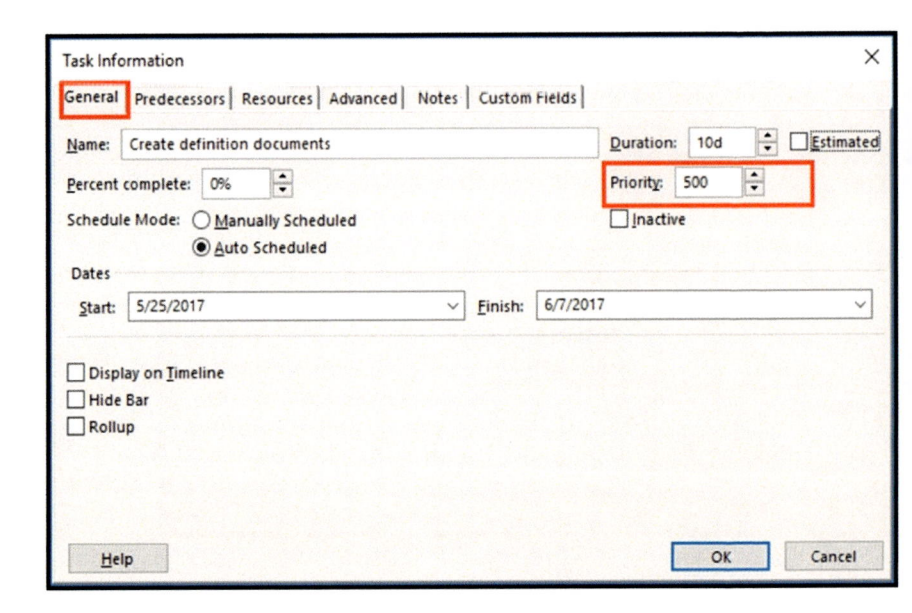

Task priority should not be random. You need to plan with the understanding that the task may be delayed or not depending on the number assigned. Since it is used in both "Standard" and "Priority, Standard" leveling, take care that your orders consider the priority of all of the tasks that are overallocated in the same period.

> Do This: If you identify a task in your schedule that must occur on its current start date, consider giving it a Priority of 1000. The task will not be delayed for leveling. You might want to make a note in the Notes tab of Task Information that the task has a high priority.

Figure 6.19: The Notes tab in the Task Information dialog can store useful information such as the reasoning for Priority changes.

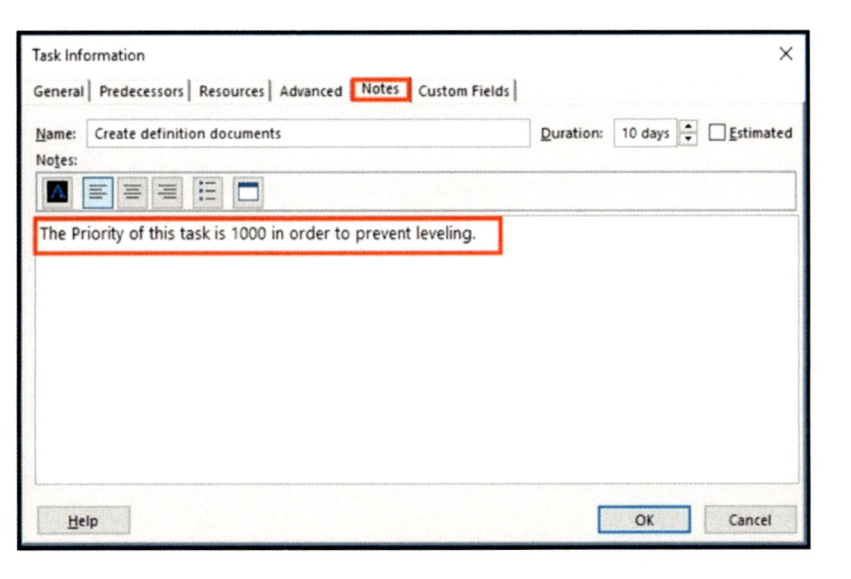

Resolving Overallocation Checkboxes

Figure 6.20: The Resource Leveling dialog's Resolving Overallocation checkboxes.

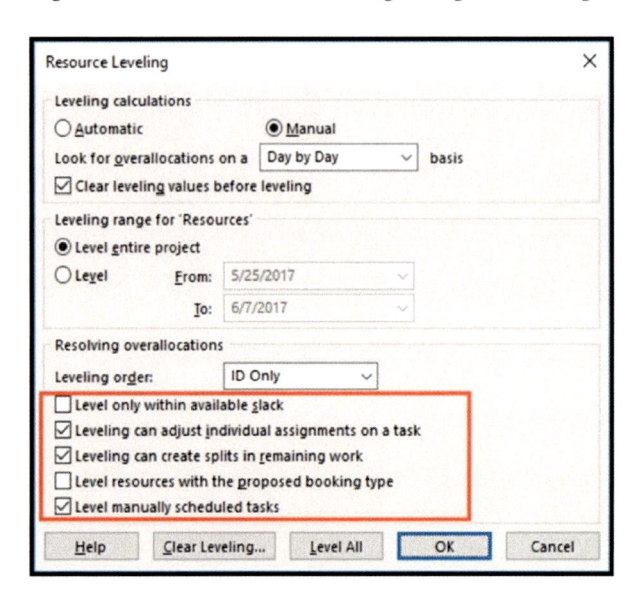

"Level only within available slack": This setting prevents the finish date of your project from being delayed. In general, it should be checked if the finish date is non-negotiable. Unchecked allows the finish date to be delayed by whatever number of working days is required to level resources. The default for this checkbox is unchecked.

"Leveling can adjust individual assignments on a task": If checked, this setting allows leveling to adjust when a resource works on a task. The work can be at a reduced number of hours or adjusted independently of other resources working on the same task. If unchecked Project will not attempt to adjust assignments. To level, Project may have to impose more delay due to its inability to modify assignments. The default is checked.

"Leveling can create splits in remaining work": This setting allows Project to stop current work and create splits in the remaining work. This is checked by default. The split will be indicated with ellipses (...) between the stop date and the resume date for the task requiring the split. If it is unchecked, Project will not split tasks for leveling purposes, which also means that more leveling delays are imposed on the task and resource schedules.

"Level resources with the proposed booking type": This setting is unchecked by default. "Proposed" resources and procedures and practices dealing with their use are an advanced topic and beyond the scope of this book. The default resource considered here is the assigned ("Committed") resource. If checked, proposed resources can be leveled. If unchecked, proposed resources will remain overallocated.

"Level manually scheduled tasks": Check this option if you want leveling to delay manually scheduled tasks that are assigned resources. If left unchecked, Project will not attempt to level *any* manually scheduled task.

> Do This: Start with the default options when you need to level resources, then modify them to meet your needs. For example, if your project finish date is not negotiable, check the "Level only within available slack" checkbox. Project will do its best to level and then give error messages telling you which tasks it cannot level because of the lack of slack. In this case, you will have to find additional resources if you expect to meet the project schedule requirements.

Tools Used in Leveling

When leveling, it is best to use tools designed to analyze and illustrate the problem, the solution and a comparison of both to take the appropriate action. One such tool is the "Leveling Gantt."

The Leveling Gantt: In this view, Project can illustrate the tasks in the schedule before they've been leveled and after. This is particularly helpful in predicting the post-leveled schedule's impact on task and project finish dates as well as milestone events.

Figure 6.21: The Leveling Gantt showing a pre- and post-leveled schedule. In this example, Project delayed task 2 by 14 calendar days to level.

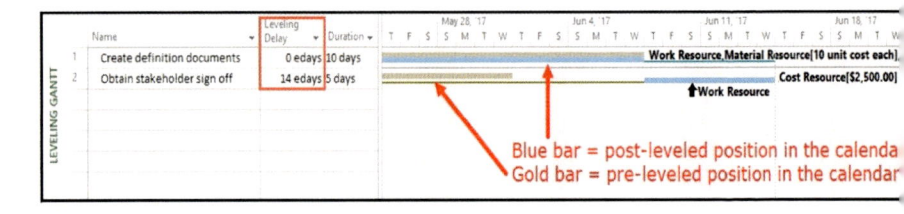

Combination Views

Two additional views can assist you in making decisions about leveling:

Resource Allocation view displays the resource usage view over the leveling Gantt.

Figure 6.22: The Resource Allocation view showing overallocation.

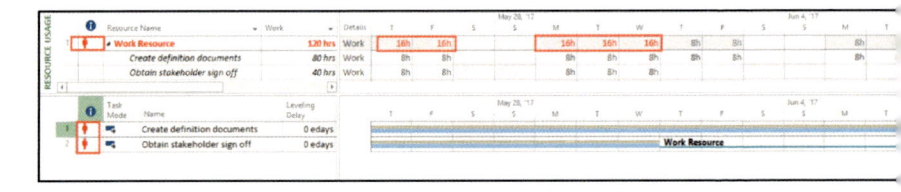

Resource Graph over the Leveling Gantt can offer visible clues and formatting of the leveling issue along with possible solutions.

Figure 6.23: The Resource graph over the Leveling Gantt, illustrating the same overallocation problem as Resource Allocation.

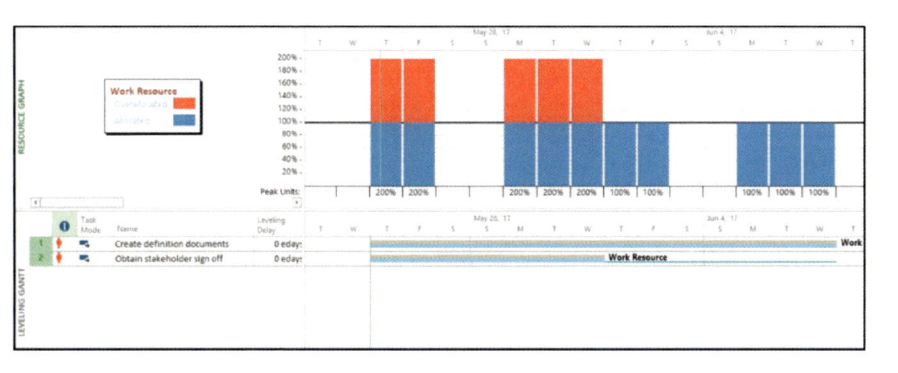

Leveling Buttons

Once you have identified who is overallocated and what is causing the problem, you can then impose your leveling tactic. Project's leveling buttons will support leveling one task or a selection of tasks, an individual resource or a group of resources or everything and everyone overallocated in your project. These leveling buttons are found in the Resource tab in the Level command group.

Figure 6-24: Leveling buttons available in the Resource tab.

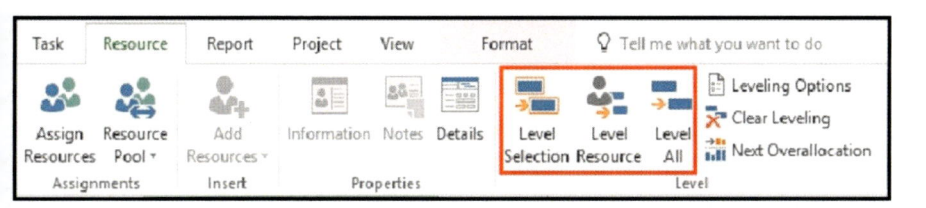

Individual tasks or groups of tasks must be selected before the Level Selection button will be active and available for use. Before leveling an individual resource or group, select the resource and click on the Level Resource button. To level the entire project, click on the Level All button.

How to Clear Leveling and Try a Different Leveling Setting

Once Project has finished leveling, doublecheck the results in the Leveling Gantt to ensure you have a workable schedule! If you don't, click on the "Clear Leveling" button and try a different approach in the Leveling Options dialog.

Project will level your project, but that does not mean that the process of leveling has created the optimum schedule. It may take several iterations of leveling using different leveling options to get the schedule you want or need.

Do's and Don'ts Review

Do's:
- Consider the project's total work estimate prior to committing resources. Is there enough work to justify the overhead of managing resources in the Project file?
- Always adjust resource Max Units to the amount of resources that will be available at any point in the project schedule.
- Check resource assignments as you create them. By catching mistakes early, you will have a better chance at creating a viable resource work schedule.
- Insert the Peak column next to the Max Units column.
- Leave your leveling calculations set to Manual.
- If you identify a task in your schedule that must occur on its current start date, consider giving it a Priority of 1000.
- Start with the default options when you need to level resources, then modify them to meet your need.

Don'ts:
- Don't define resources in views other than the resource sheet.
- Don't assign resources to summary tasks.

Chapter 7: Baseline Your Project

In this chapter, you will learn:

- The benefits of baselines
- Methods of setting a baseline
- The concept of variance
- How to interpret variance

Benefits of a Baseline

A baseline is a static snapshot of your schedule, your costs and work. It is far more than just a reporting requirement! Having a baseline and being able to track project performance against it means that you can see when performance is lagging, when costs are rising and when work is slowing. You can make schedule, cost and work predictions that have a chance of being true.

It gives you and your organization the ability to save the history of the project in order to learn and improve project management capabilities. Without a baseline to measure progress against, reporting performance is reduced to guesswork.

Methods of Setting a Baseline

There is some confusion as to when a baseline needs to be captured and saved. One opinion is that the entire project baseline needs to be set prior to any work being done. Another is to baseline in stages. Both are correct if your organization supports one over the other. I am going to discuss two options: 1) baselining the entire project; and 2) saving partial baselines to accommodate new tasks and specific groups of tasks in the project.

Baselining the Entire Project

Baselines are set in the Set Baseline dialog. This dialog is available by clicking on the Microsoft Project tab and then on Set Baseline in the Schedule command group. Once you have clicked on the Set Baseline button, Project will offer you two choices: Set Baseline... and Clear Baseline.

Click on "Set Baseline..."

Figure 7.1: The choices in "Set Baseline."

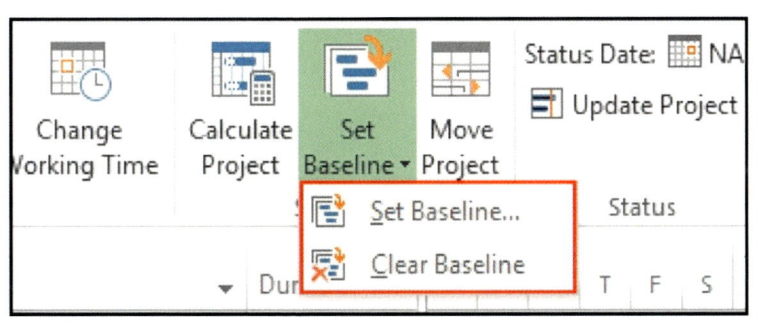

The "Set Baseline" dialog appears and offers you the default of setting the baseline for the entire project. If you click on the OK button, the entire project is baselined.

Figure 7.2: The Set Baseline dialog.

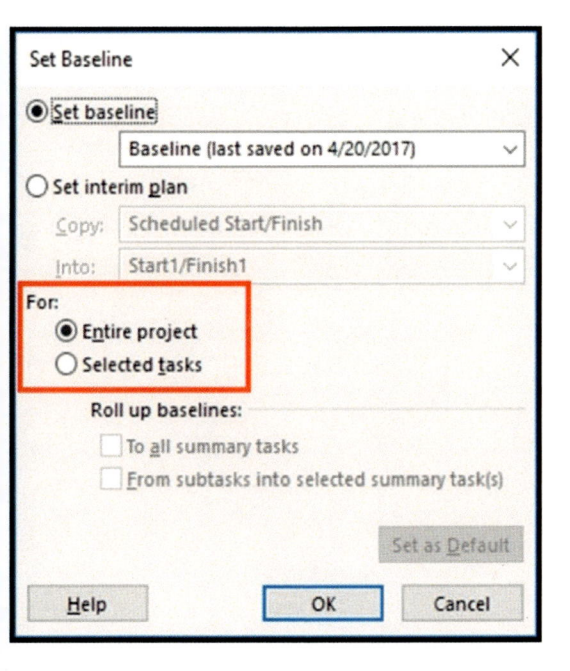

Ch 7: Baseline Your Project | **97**

Once the baseline is set, the Tracking Gantt view lets you see the current schedule compared to the baseline schedule. Since this view updates as data is entered, the comparison is as fresh as your data. It will also show you any differences in the current task dates when compared to the baseline dates.

Figure 7.3: The Tracking Gantt compares the current schedule to the baseline.

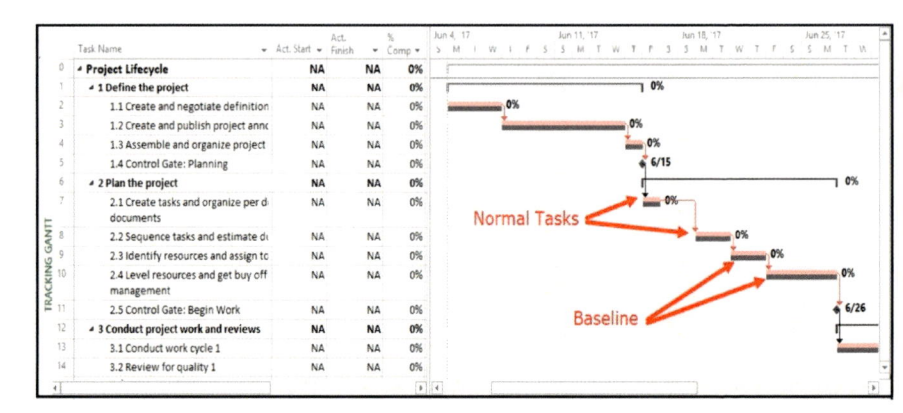

When differences begin to appear in the schedule, those differences are called "variances." If the start date changes on a task, the current task bar will move in the timeline to match the new start date. The baseline will not move, thereby allowing you to make comparisons and reach decisions on correcting lagging tasks.

Figure 7.4: The Tracking Gantt showing Start Variance.

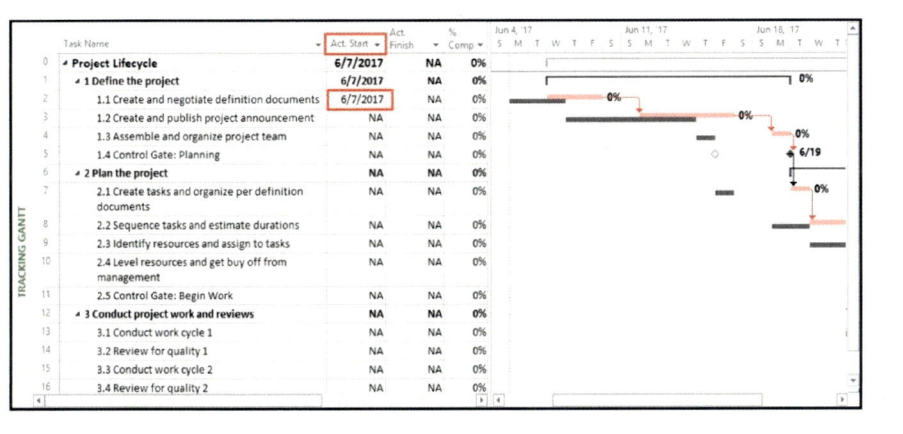

Baselining Individual Tasks or Groups of Tasks

When new tasks are added to a project, they have no baseline. It is important that they are given a baseline; otherwise tracking performance on them will never show schedule variance to "Start," "Duration" or "Finish."

It is also important not to overwrite the pre-existing baseline with a new baseline. Variance may already exist in the schedule, and if the pre-existing baseline is overwritten, the variance values would be lost. This overwriting occurs when you just re-baseline the entire project without checking whether you should. If the baseline already exists and variance has already occurred, the Tracking Gantt will show you the schedule impact of the variance. Project will then calculate a new project finish date if necessary. Each time you set the baseline for the entire project, the variance is reset to zero.

> **Do This:** When setting a baseline for new tasks, ensure that the existing baseline is not overwritten!

To set a baseline on new tasks only, first select the tasks in the task list, then choose the Project tab. Next, in the Schedule command group click on Set Baseline. Now Project will offer you the choices, Set Baseline… and Clear Baseline… Click on Set Baseline…

The "Set Baseline" dialog appears and offers you the default of setting the baseline for the entire project. In the "Set Baseline" dialog you will need to change the For: default from Entire Project to Selected tasks. Next, ensure the baseline data rolls up through summary tasks by checking the boxes in Roll up baselines:.

Figure 7.5: The Set Baseline dialog with the selected tasks and roll up options set.

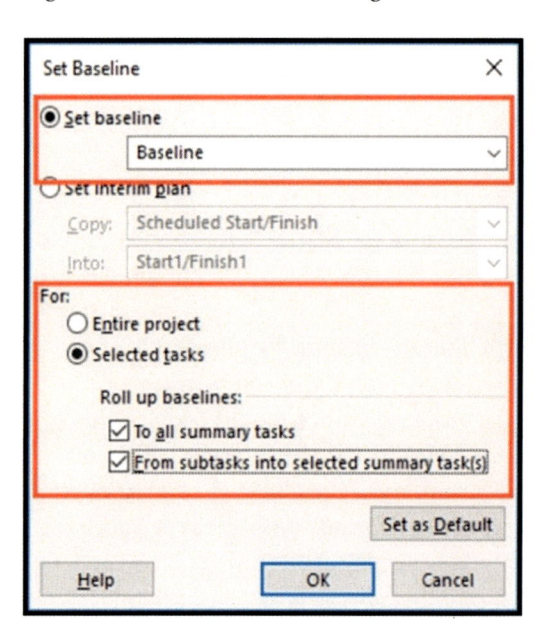

Once these changes are made, the baseline for the new tasks will be set after a warning that the baseline has been set already. Since we are baselining the selected tasks and not the entire project, it is safe to dismiss the warning by clicking on the "Yes" button.

Figure 7.6: The Set Baseline warning. This warning only occurs when a baseline already exists in the project.

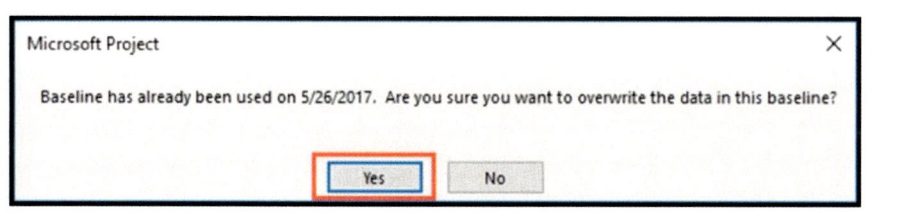

When the last task is baselined, you are ready to track progress and show variances. To track accurately, a baseline should be set on tasks before tracking values are entered.

> **Do This:** Set your baseline before entering actuals. Accurate variance calculations require it!

The Concept of Variance

Variance just means that there is a difference between what was planned and what is happening. Variance is calculated by Project when you enter progress and the results of the calculations are entered into Project's variance fields.

If a task starts two days late, it will display two days of Start Variance. If it starts two days early, it will have minus-two days in Start Variance. Finish dates and durations operate similarly: "Early" will be represented by a negative value and "later" by a positive.

> **Don't Do This:** Don't take a "wait and see" approach to variance! Variance seldom self-corrects and the result of variance may be to miss your schedule goals. When variance is observed, make sure it is recorded, communicated and managed.

Figure 7.7: A Tracking Gantt and Variance table showing positive and negative variance.

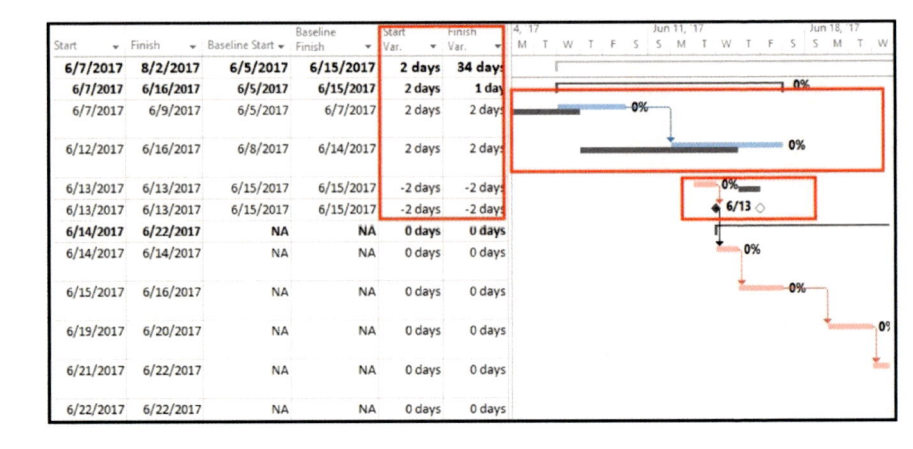

How to View Variance

The Tracking Gantt is an excellent view to examine task start and finish variances graphically. The comparison of the baseline and current task bar in this type of Gantt offers a powerful at-a-glance status report. But the graphics need to be backed up with the calculated variance fields to communicate the severity of the variance. Project's tables give you specific types of variance in logical tables.

Variance table: This type of logical table shows start and finish variance. Negative numbers mean the task Start or Finish is before the Baseline Start or Finish.

Figure 7.8: The Variance table.

Work table: This type of table shows work variance. Negative numbers mean that the Work is less than the Baseline work.

Figure 7.9: A Work table showing work variance.

	Task Name	Work	Baseline	Variance	Actual	Remaining
0	▲ Project Lifecycle	1,396 hrs	392 hrs	1,004 hrs	176 hrs	1,220 hrs
1	▲ 1 Define the project	176 hrs	176 hrs	0 hrs	176 hrs	0 hrs
2	1.1 Create and negotiate definition documents	72 hrs	72 hrs	0 hrs	72 hrs	0 hrs
3	1.2 Create and publish project announcemer	80 hrs	80 hrs	0 hrs	80 hrs	0 hrs
4	1.3 Assemble and organize project team	24 hrs	24 hrs	0 hrs	24 hrs	0 hrs
5	1.4 Control Gate: Planning	0 hrs	0 hrs	0 hrs	0 hrs	0 hrs
6	▲ 2 Plan the project	260 hrs	216 hrs	44 hrs	0 hrs	260 hrs
7	2.1 Create tasks and organize per definition documents	40 hrs	40 hrs	0 hrs	0 hrs	40 hrs
8	2.2 Sequence tasks and estimate durations	100 hrs	80 hrs	20 hrs	0 hrs	100 hrs
9	2.3 Identify resources and assign to tasks	80 hrs	64 hrs	16 hrs	0 hrs	80 hrs
10	2.4 Level resources and get buy off from management	40 hrs	32 hrs	8 hrs	0 hrs	40 hrs
11	2.5 Control Gate: Begin Work	0 hrs	0 hrs	0 hrs	0 hrs	0 hrs

Cost table: This one displays cost variance. Negative numbers mean the Total Cost is less than the Baseline cost.

Figure 7.10: A Cost table showing cost variance.

	Task Name	Fixed Cost	Fixed Cost Accrual	Total Cost	Baseline	Variance	Actual
0	▲ Project Lifecycle	$0.00	Prorated	$121,580.00	$42,648.00	$78,932.00	21,248.00
1	▲ 1 Define the project	$0.00	Prorated	$21,248.00	$21,248.00	$0.00	$21,248.00
2	1.1 Create and negotiate definition documents	$0.00	Prorated	$8,400.00	$8,400.00	$0.00	$8,400.00
3	1.2 Create and publish proje	$0.00	Prorated	$10,048.00	$10,048.00	$0.00	$10,048.00
4	1.3 Assemble and organize	$0.00	Prorated	$2,800.00	$2,800.00	$0.00	$2,800.00
5	1.4 Control Gate: Planning	$0.00	Prorated	$0.00	$0.00	$0.00	$0.00
6	▲ 2 Plan the project	$0.00	Prorated	$25,900.00	$21,400.00	$4,500.00	$0.00
7	2.1 Create tasks and organize per definition	$0.00	Prorated	$3,400.00	$3,400.00	$0.00	$0.00
8	2.2 Sequence tasks and esti	$0.00	Prorated	$7,500.00	$6,000.00	$1,500.00	$0.00
9	2.3 Identify resources and a	$0.00	Prorated	$10,000.00	$8,000.00	$2,000.00	$0.00
10	2.4 Level resources and get buy off from management	$0.00	Prorated	$5,000.00	$4,000.00	$1,000.00	$0.00
11	2.5 Control Gate: Begin Wor	$0.00	Prorated	$0.00	$0.00	$0.00	$0.00

To view project level variance as an overview, call up "Project Statistics" from the Project Information dialog.

Figure 7.11: The Project Information dialog. The Project Statistics button is at the bottom of the dialog.

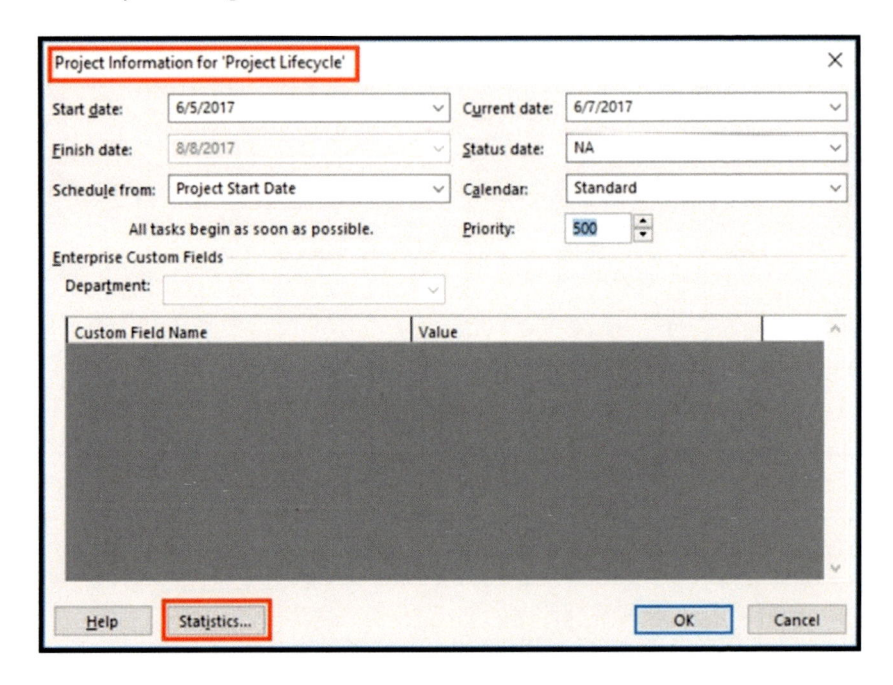

Figure 7.12: Project Statistics showing variances.

<u>Do This:</u> Once the baseline is in place and tracking progress has begun, view statistics often! You will get your first clue that variance is occurring and have a better chance of reacting to it and reporting on it.

Do's and Don'ts Review

When a baseline is set by a project manager, it generally means the project is now ready to start. Understanding variance and getting it in control is a fundamental skill required of the project manager.

Do's:
- Set your baseline before entering actuals. Accurate variance calculations require it!
- When setting a baseline for new tasks, don't allow the existing baseline to be overwritten!
- Once the baseline is in place and tracking progress has begun, view statistics often!

Don'ts:
- Don't take a "wait and see" approach to variance. When variance is observed, make sure it is recorded, communicated and managed.

Chapter 8: Track the Project

In this chapter, you will learn:

- Why to track a project
- Basic tracking methods
- How to respond to the realities of work that finishes early or late

Reasons for Tracking Your Project

Projects almost never go as planned. At best, projects represent refined guesses about work to be done into the future. These are easily undermined by overly optimistic schedules, inaccurate work and cost estimates, inadequate funding, inadequate staffing and any number of natural disasters.

Tracking helps us get better at our projects by managing realistically and transparently as possible. This means using baselines for comparisons, performing frequent Microsoft Project file updates and being consistent in what is tracked by following a "tracking method."

Work, cost and schedule are the typical variables tracked in Project. Progress can be reported at the level you find convenient: task, summary or project. Reporting will be covered in more detail in the next chapter.

> Do This: Be consistent! Decide on a tracking interval supporting the reporting frequency. If you must report every week, make sure you update your project information weekly and have a chance to consider changes before reporting. Being consistent gives you the possibility of a systematic and repeatable method of gathering, considering and reporting project progress.

> Don't Do This: Don't track at intervals longer than bi-weekly. Problems tend to get worse the longer they are ignored or unseen. Having a shorter interval between tracking sessions means you'll find problems earlier and have a better chance of resolving them.

Tracking Definitions

Figure 8.1: The Tracking table contains the fields necessary to track progress.

Task Name	Act. Start	Act. Finish	% Comp.	Phys. % Comp.	Act. Dur.	Rem. Dur.	Act. Cost	Act. Work
⏷ Project Lifecycle	6/5/2017	NA	40%	0%	19.38 days	29.13 days	$51,573.00	515 hrs
⏷ 1 Define the project	6/5/2017	6/15/2017	100%	0%	9 days	0 days	$21,248.00	176 hrs
1.1 Create and negotiate definitic	6/5/2017	6/7/2017	100%	0%	3 days	0 days	$8,400.00	72 hrs
1.2 Create and publish project ani	6/8/2017	6/14/2017	100%	0%	5 days	0 days	$10,048.00	80 hrs
1.3 Assemble and organize projec	6/15/2017	6/15/2017	100%	0%	1 day	0 days	$2,800.00	24 hrs
1.4 Control Gate: Planning	6/15/2017	6/15/2017	100%	0%	0 days	0 days	$0.00	0 hrs
⏷ 2 Plan the project	6/16/2017	NA	75%	0%	6.38 days	2.13 days	$19,425.00	195 hrs
2.1 Create tasks and organize per	6/16/2017	NA	75%	0%	0.75 days	0.25 days	$2,550.00	30 hrs
2.2 Sequence tasks and estimate (	6/19/2017	NA	75%	0%	1.88 days	0.63 days	$5,625.00	75 hrs
2.3 Identify resources and assign 1	6/21/2017	NA	75%	0%	1.88 days	0.63 days	$7,500.00	60 hrs
2.4 Level resources and get buy of	6/26/2017	NA	75%	0%	1.88 days	0.63 days	$3,750.00	30 hrs
2.5 Control Gate: Begin Work	6/28/2017	NA	75%	0%	0 days	0 days	$0.00	0 hrs
⏷ 3 Conduct project work and reviews	6/28/2017	NA	15%	0%	4 days	22 days	$10,900.00	144 hrs
3.1 Conduct work cycle 1	6/28/2017	NA	50%	0%	3.5 days	3.5 days	$10,500.00	140 hrs
3.2 Review for quality 1	7/7/2017	NA	50%	0%	0.5 days	0.5 days	$400.00	4 hrs
3.3 Conduct work cycle 2	NA	NA	0%	0%	0 days	7 days	$0.00	0 hrs
3.4 Review for quality 2	NA	NA	0%	0%	0 days	1 day	$0.00	0 hrs
3.5 Conduct cycle work 3	NA	NA	0%	0%	0 days	7 days	$0.00	0 hrs
3.6 Review for quality 3	NA	NA	0%	0%	0 days	3 days	$0.00	0 hrs
3.7 Control Gate: Close and Docur	NA	NA	0%	0%	0 days	0 days	$0.00	0 hrs

To understand progress in Project, you need to understand the primary fields used in tracking:

Actual Start: The date that work begins in real life. Actual Start is a column found in the Tracking table.

Actual Finish: The date that work is completed. Actual Finish is a column found in the Tracking table.

Actual Work: The amount of work that has been completed by resources assigned the task. Actual Work is a column found in the Tracking and Work tables.

Actual Cost: The accrued cost of work already performed by resources assigned the task plus other costs. Actual Cost is a column found in the Tracking and Cost tables.

Actual Duration: The amount of task schedule completed so far. Actual Duration is a column found in the Tracking table. The formula is:

*Duration * % of tasks completed*

For example, if a five-day task is 60% complete, its actual duration is three days. Alternatively, you could state that 60% of the task's allotted schedule is complete.

Remaining Work: The amount of work remaining in a task. The formula is:

Total work - Actual work completed

If the five-day task had 40 hours of effort scheduled to be completed and 24 hours of effort actually *was* completed, then the Remaining Work is 16 hours.

Remaining Duration: The amount of time left in the schedule. The formula is:

Duration - Actual Duration

Using the five-day task as an example again, if the Actual Duration is three days, then the five days' Duration minus three days' Actual Duration equals two days' Remaining Duration. Alternatively, you can state that 40% of the task's allotted schedule remains to be completed.

%Complete: Progress based on how much of a task's duration has been completed. The formula is:

*(Actual Duration / Duration) * 100*

Using the same example as above, three days' Actual Duration / five days' Duration results in 0.60, multiplied by 100 results in 60% Complete.

%Work Complete: Progress based on how much work has been completed on a task. The formula is:

*(Actual Work / Work) * 100*

In this final example, 24 hours of actual work / 40 hours total work in the task results in 0.60, multiplied by 100 results in 60% Work Complete. In summary, the status of our five-day task is:

- Work and schedule progress is 60% complete;
- 24 hours of work have been completed;
- 16 hours of work remain; and
- Two days of duration remain to conduct the remaining work.

Common Methods Used in Tracking

Tracking can be tedious. It's more tedious when you don't know the methods that are suited to you and your organization's needs. What follows is the information you need to be able to select a *situational* tracking method—one that fits your need *when you need it*.

Mark on Track: If a task started on time, is underway and no problems have appeared yet, updating it in the project plan is easy. Just click on the task, select the Task tab and then in the Schedule command group click on Mark on Track. Project will calculate the completion to date, and you can then move on to other tasks needing your time and attention.

Figure 8.2: The Mark on Track button.

Update Tasks: When tasks start early or late, Project will need more information to track accurately. The actual dates the task started and finished, actual and remaining duration and %complete may all need updating. The Update Tasks dialog is a handy tool you might use to enter the tracking variables in one place. The Update Tasks button and dialog is found in the Mark on Track dropdown list.

Figure 8.3: Mark on Track and Update Tasks buttons.

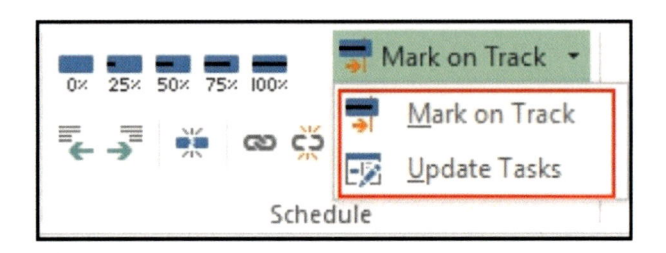

Figure 8.4: The Update Tasks dialog.

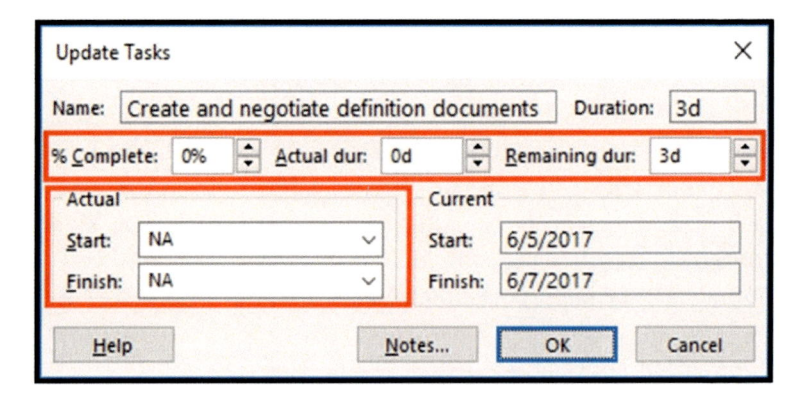

✓ Do This: Any time a task has started, make sure you enter its "Actual Start." This ensures that progress can be applied from its actual start. With an Actual Start date applied to a task, Project knows that the task schedule *must* begin there. If the task actual finish date is different than its scheduled finish, enter the actual finish date. Project will then make the calculations to work and cost that represent the actual start and finish dates.

When accurate, actual tracking data is collected and compared to the baseline data, the team and the organization have historical information that can be used the next time a similar task or project is created. Over time, estimating accuracy should improve as tracking history is documented, learning occurs and you have the chance to get better at project execution.

%Complete: When you lack administrative time to spend on entering completion values, consider using the tracking buttons in the Task tab and Schedule command group. These buttons will help you set the schedule and work completion to the same value. So, if you selected a task and then pressed the "25%" button, the schedule completion would be set to 25% complete and the %Work Complete would be set to 25% as well.

Figure 8.5: *The tracking buttons set schedule and work completion to the same percentage.*

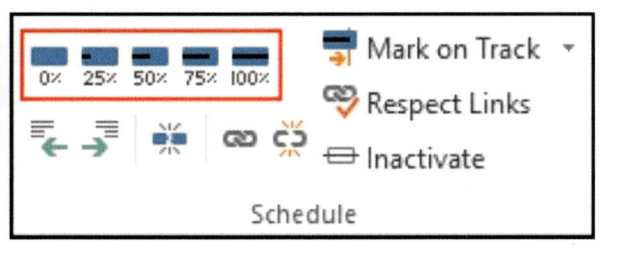

The %Complete buttons are very useful if a simplified tracking method is required. Often project managers use only these values in tracking schedule completion:

- 0% for not started;
- 25%, 50% and 75% for started but not finished; and
- 100% for completed.

Naturally, the values can be even simpler: 0%, 50% and 100%, for example.

> **Don't Do This:** Don't track into the future! It's very easy to get lulled into using only the %Complete buttons. Look first at the task's Start Date. If the date is in the future, any completion means it really started early. You need to give Project the Actual Start date. Otherwise, Project will calculate from the task's scheduled start in the future!

Actual and Remaining Work

When accuracy of the status is paramount, %Complete may be too arbitrary for your organization's need. In that case, you will want to track the actual work that has been finished to date and estimate the amount of work that remains to complete the task. These completed entries then calculate a "%Complete" using verified data and not an arbitrary button.

While this method takes more administration time and more communication with the team, it gives you the confidence that you have done your best to validate true task schedule and work status.

An easy approach to this manner of tracking is to view the Tracking Gantt, then click on the Details checkbox found in the Split View command group in the View tab. This action changes the screen to a combination view, with the Tracking Gantt in the Primary or top window and a Task form in the Details or lower window.

Right-click on the task form to see a list of other formats available for that. Clicking on Work will change the form content so that Actual Work and Remaining Work may be entered on a task-by-task basis in your task list. When the form is OK'd, the task %Complete will be calculated based on your entries in the task form.

Figure 8.6: Tracking Gantt over the Task form showing the Work detail.

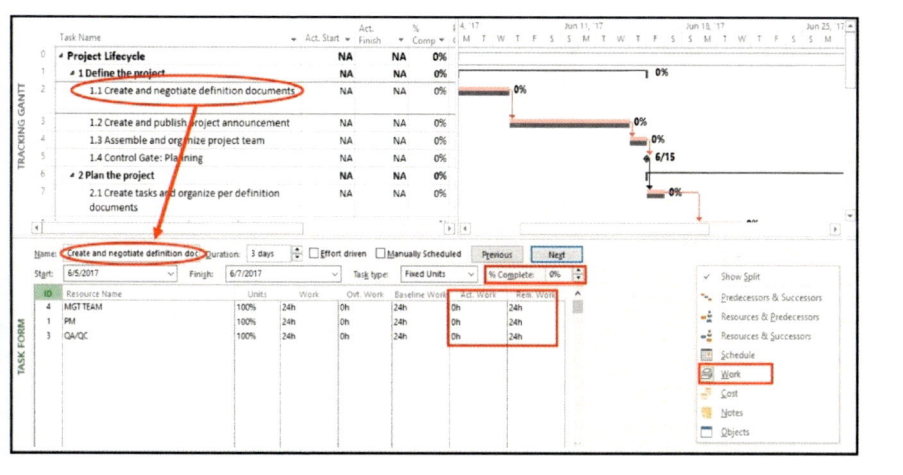

Move Commands Aid Rescheduling Work

Sometimes the work does not go well. Tasks don't start on time, people may be pulled from your project to work on higher priority activities or work just stalls. That's when it is time to pull out your backup plan.

Fortunately, Project was designed with backup tactics in mind. For the project manager, this means the ability to manage the work and schedule interruptions by rescheduling work using the Move commands.

Move is found in the Task tab in the Tasks command group.

Figure 8.7: The Move button.

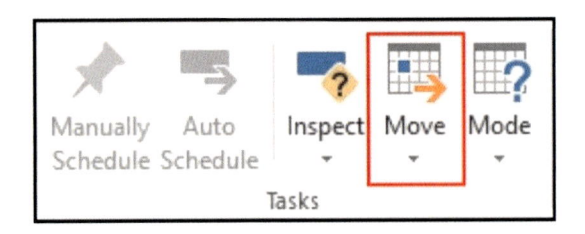

A dropdown list under the Move button provides many options designed for re-planning tasks whose schedule and work has fallen behind.

Figure 8.8: Move dropdown list options.

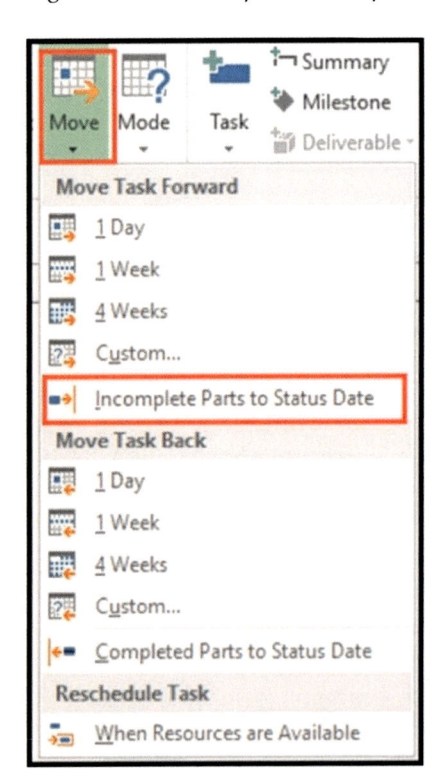

Tasks can be moved forward or back in the timeline by a prescribed amount of time or an amount of time you choose in the "Custom..." dialog boxes.

When work stalls and needs rescheduling, you can move the remaining work to the current date or a status date by choosing "Incomplete Parts to Status Date." If work never started, the entire task will be moved.

Work sometimes starts earlier than expected. When that happens, you will need to edit the task information in Update Tasks and use the Move command for Project to represent the status in the schedule. If the task has predecessors, make sure that the link will not interfere with moving the task. This is particularly true in the instance of a task that starts early. You may have to remove the predecessor link so that Project can schedule the task and its successors correctly.

> Do This: When moving tasks earlier or later in the timeline, pay close attention to the task's predecessors and successors. Moving a task carelessly can create other schedule, work and cost issues!

Once the task is in the correct schedule position and you are satisfied with it, you can track the task and project through its completion.

Do's and Don'ts Review

Do's:
- Be consistent! Decide on a tracking interval supporting the reporting frequency.
- Any time a task has started, make sure you enter its "Actual Start." When the finish date is not what you expected, enter the "Actual Finish."
- When moving tasks earlier or later in the timeline, pay close attention to the task's predecessors and successors.

Don'ts:
- Don't track at intervals longer than bi-weekly.
- Don't track into the future! Enter the actual start if a task is started early!

Chapter 9: Reporting

In this chapter, you will learn:

- Project's reports
- Visual reports
- Formatting views as reports

Project's Reports

Since the release of Microsoft Project 2013, Project's reports have been internally generated, customizable, printable and savable (a valuable departure from Project 2010 and earlier versions). You will find all of your reporting options on the Report tab. Your options are in the View Reports and Export command groups. There are many reports available, and the ability to customize them is limited only by imagination. There are a few that stand out and deserve more comment here.

Figure 9.1: The Report tab and its command groups.

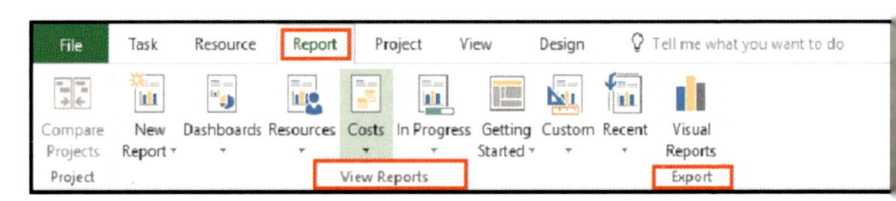

Dashboards reports: Summary reports and charts that represent your data in a filtered, grouped and structured manner. You can see this in action in the Work Overview dashboard.

In the Work Overview dashboard work is shown as a diminishing value in a "Burndown" chart. This chart compares the amount of work done to the amount of work remaining over the project duration.

The dashboard also contains a Work Stats diagram, comparing Actual Work, Remaining Work and Baseline Work at the summary task level.

Figure 9.2: The Work Overview dashboard Work Burndown and Work Stats charts.

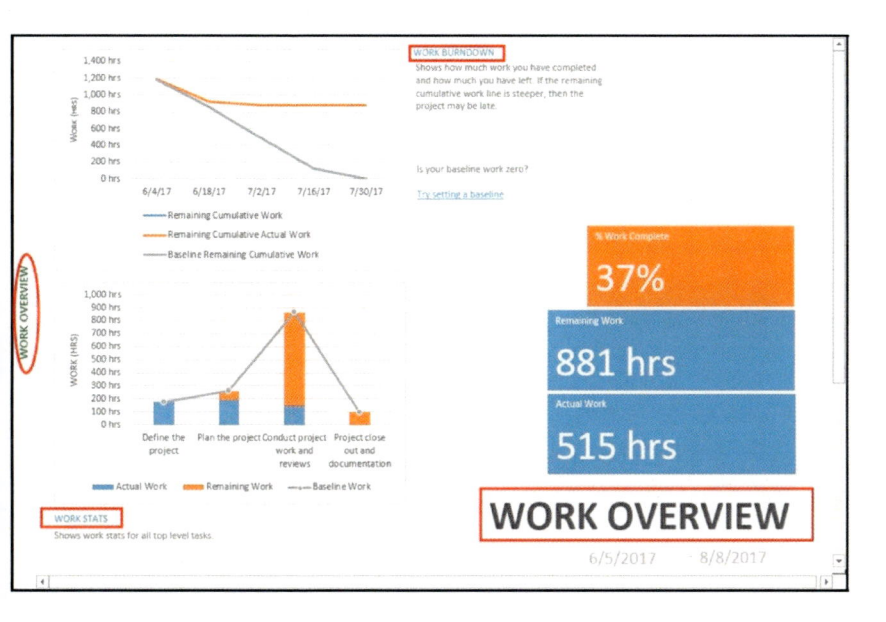

Finally, the dashboard compares resource actual and remaining work in a Resource Stats chart and work and availability of resources in a Remaining Availability chart. You have a clear picture of all work in the project. There is a tremendous amount of information summarized for data-informed decision-making.

Figure 9.3: The Work Overview dashboard with Resource Stats and Remaining Availability charts.

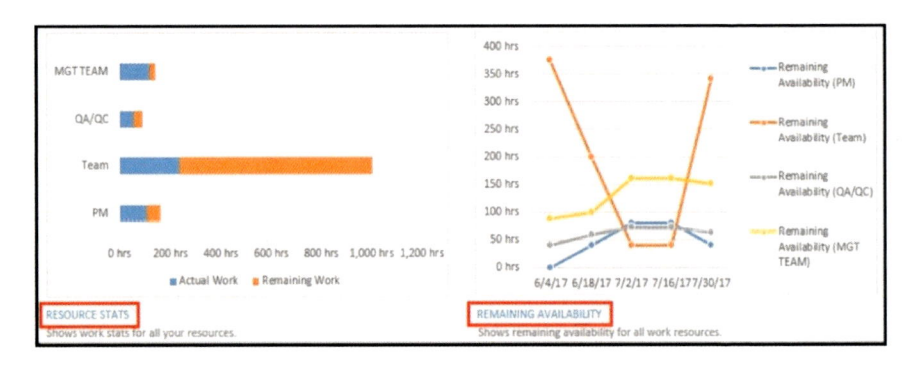

Click on any of the charts, graphs tables and diagrams in Project's View Reports and Project will offer you Report Tools to design your report, Chart Tools to format it and a Field List in a sidebar in case you need to customize the pre-defined reports.

Figure 9.4: The Work Overview dashboard with the Work Stats chart selected and the Field List sidebar visible.

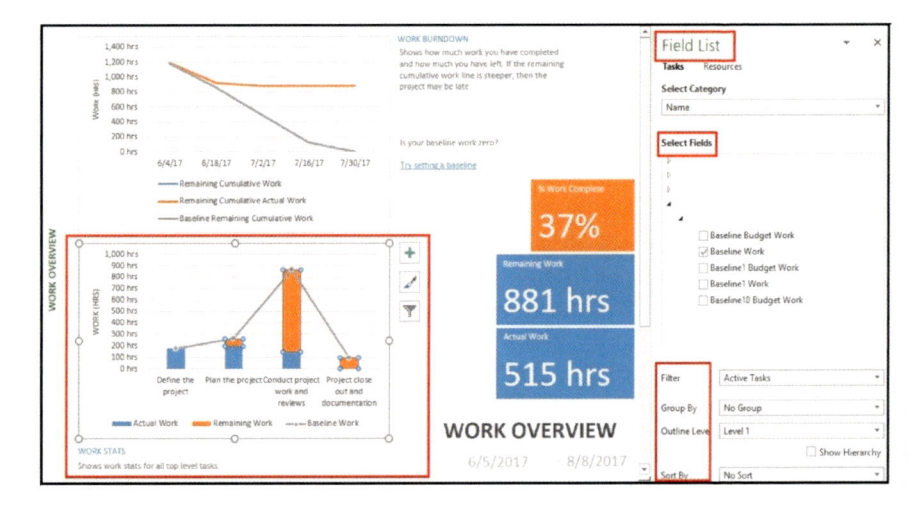

✅ <u>Do This:</u> Build your own reports. You know what information your customer, manager or executive needs, and Project report tools are flexible and forgiving.

From a simple table to complex comparison charts it is a matter of pointing, clicking and dragging to get the information you need in the format you want.

Resources report: Contains two built-in reports regarding resource work and resource over-allocation.

Resource Overview report: Summarizes resource work and progress in three charts:

Resource Stats report: Compares the actual, baseline and remaining work.

Work Status report: Shows the percentage of work completed by each resource.

Resource Status report: Identifies each resource's start and finish date on the project and how much of his or her work remains.

Figure 9.5: A Resource Overview report showing Resource Stats, Work Status and Resource Status charts.

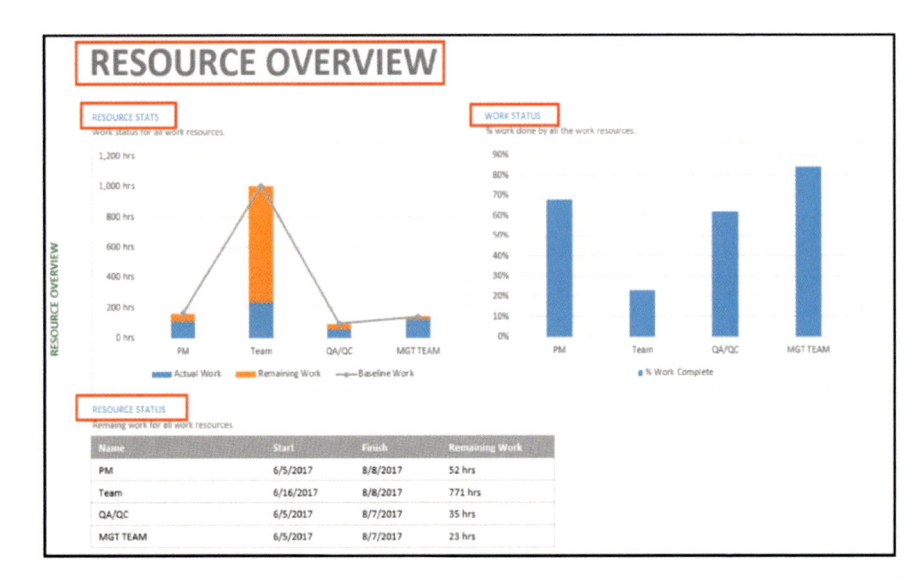

Overallocated Resources report: Compares Actual Work to Remaining Work and reports on how much work is overallocated in the project's time period. This report is a high-level overview, not a detailed report on overallocation.

Figure 9.6: The Overallocated Resources report.

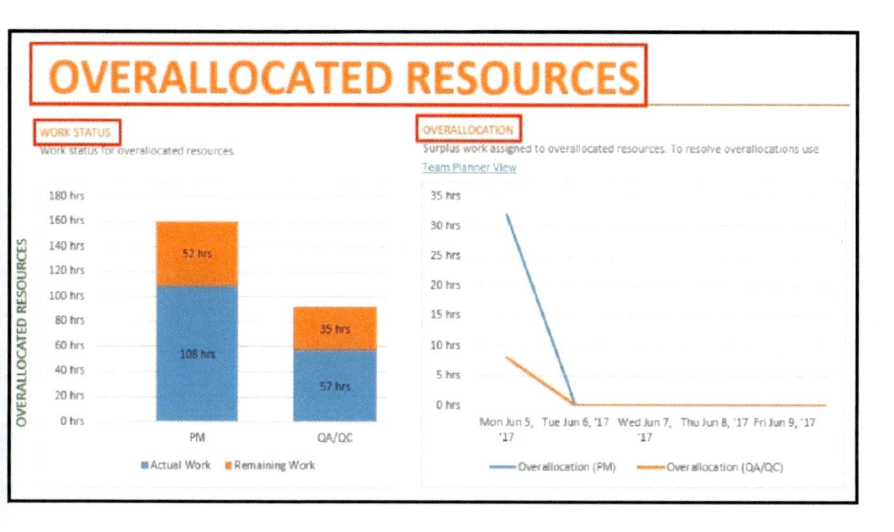

Cash Flow report: Under the Costs tab of Report is the Cash Flow report. It offers information on costs by period and shows how the costs have accrued over the duration of the project. This report can be viewed at various outline levels and timescales to meet your needs.

Figure 9.7: The Cash Flow report includes earned value calculations.

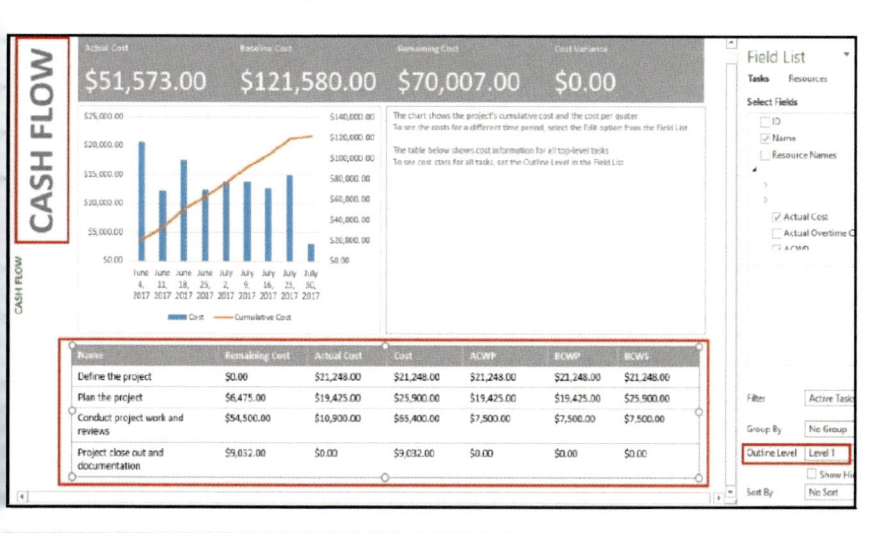

Ch 9: Reporting | 125

Resource Cost Overview report: Complements the Resource Overview report in that the cost of the resource is the focus of the report. Both resource work and cost are compared along with a table explaining the cost basis for each resource.

Figure 9.8: The Resource Cost Overview report.

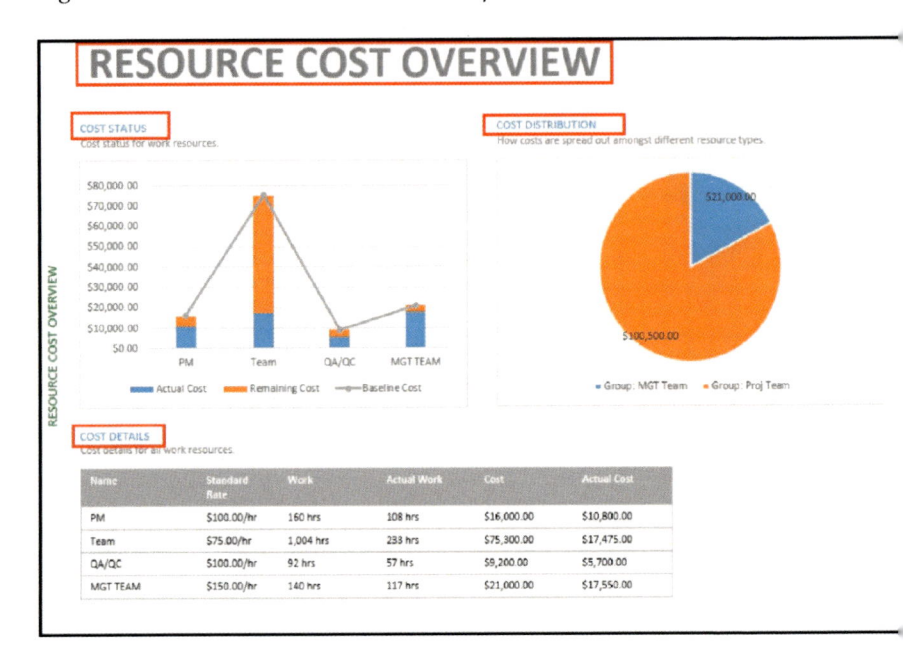

In Progress report: Provides indispensable information on progress once work begins and assists you in identifying potential and real problems in your schedule.

Critical Task report: Identifies critical tasks—those tasks that have a schedule that cannot be delayed any further.

Figure 9.9: The Critical Tasks report.

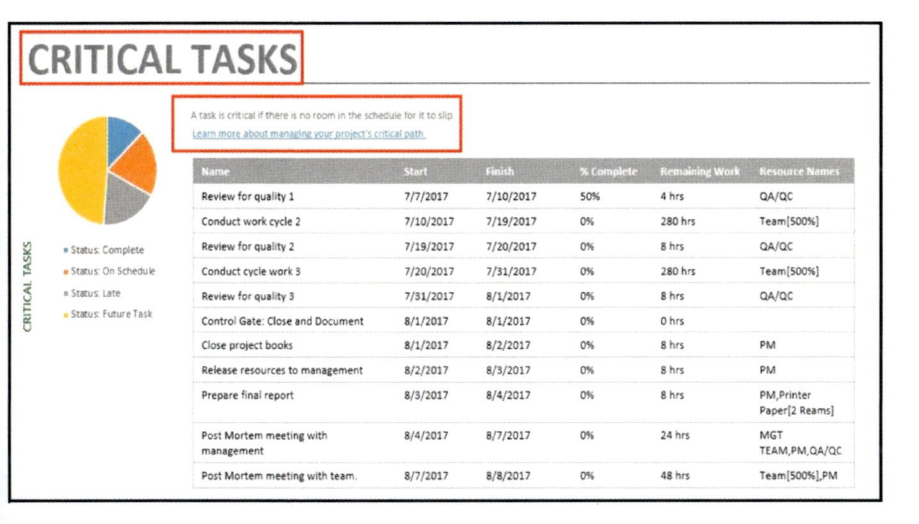

CRITICAL TASKS

A task is critical if there is no room in the schedule for it to slip.
Learn more about managing your project's critical path.

Name	Start	Finish	% Complete	Remaining Work	Resource Names
Review for quality 1	7/7/2017	7/10/2017	50%	4 hrs	QA/QC
Conduct work cycle 2	7/10/2017	7/19/2017	0%	280 hrs	Team[500%]
Review for quality 2	7/19/2017	7/20/2017	0%	8 hrs	QA/QC
Conduct cycle work 3	7/20/2017	7/31/2017	0%	280 hrs	Team[500%]
Review for quality 3	7/31/2017	8/1/2017	0%	8 hrs	QA/QC
Control Gate: Close and Document	8/1/2017	8/1/2017	0%	0 hrs	
Close project books	8/1/2017	8/2/2017	0%	8 hrs	PM
Release resources to management	8/2/2017	8/3/2017	0%	8 hrs	PM
Prepare final report	8/3/2017	8/4/2017	0%	8 hrs	PM,Printer Paper[2 Reams]
Post Mortem meeting with management	8/4/2017	8/7/2017	0%	24 hrs	MGT TEAM,PM,QA/QC
Post Mortem meeting with team.	8/7/2017	8/8/2017	0%	48 hrs	Team[500%],PM

Pie chart legend: Status: Complete, Status: On Schedule, Status: Late, Status: Future Task

Late Tasks report: Identifies tasks that are late when compared to the current date or a status date. In Project, a task is late when its finish date has passed and it is not 100-percent complete.

Figure 9.10: The Late Tasks report.

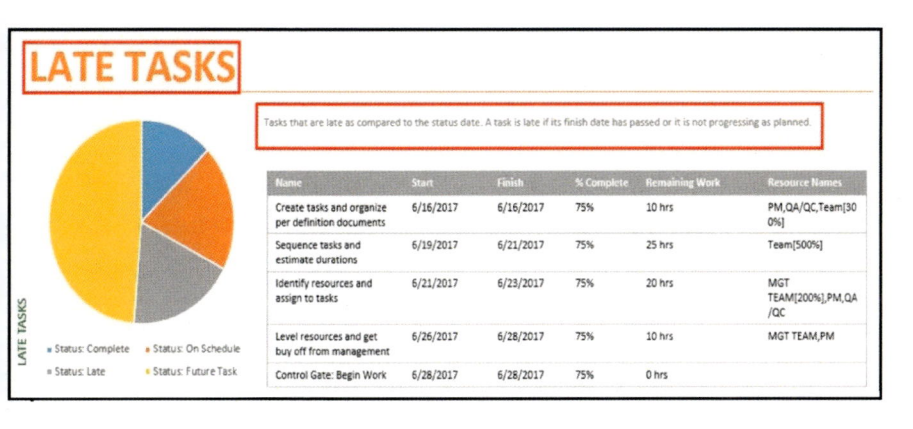

LATE TASKS

Tasks that are late as compared to the status date. A task is late if its finish date has passed or it is not progressing as planned.

Name	Start	Finish	% Complete	Remaining Work	Resource Names
Create tasks and organize per definition documents	6/16/2017	6/16/2017	75%	10 hrs	PM,QA/QC,Team[300%]
Sequence tasks and estimate durations	6/19/2017	6/21/2017	75%	25 hrs	Team[500%]
Identify resources and assign to tasks	6/21/2017	6/23/2017	75%	20 hrs	MGT TEAM[200%],PM,QA/QC
Level resources and get buy off from management	6/26/2017	6/28/2017	75%	10 hrs	MGT TEAM,PM
Control Gate: Begin Work	6/28/2017	6/28/2017	75%	0 hrs	

Pie chart legend: Status: Complete, Status: On Schedule, Status: Late, Status: Future Task

Slipping Tasks report: Identifies tasks that have a finish date later than their "Baseline Finish." The tasks may not be late yet, but their current state of progress indicates they will be late if no action is taken.

Figure 9.11: The Slipping Tasks report.

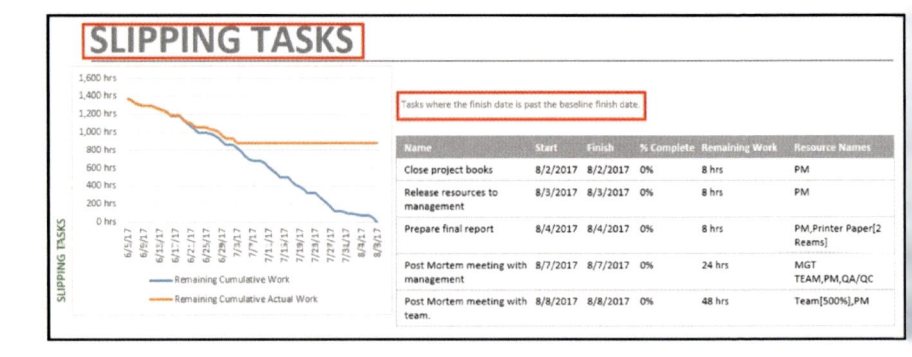

Name	Start	Finish	% Complete	Remaining Work	Resource Names
Close project books	8/2/2017	8/2/2017	0%	8 hrs	PM
Release resources to management	8/3/2017	8/3/2017	0%	8 hrs	PM
Prepare final report	8/4/2017	8/4/2017	0%	8 hrs	PM,Printer Paper[2 Reams]
Post Mortem meeting with management	8/7/2017	8/7/2017	0%	24 hrs	MGT TEAM,PM,QA/QC
Post Mortem meeting with team.	8/8/2017	8/8/2017	0%	48 hrs	Team[500%],PM

> **Do This:** Use the Slipping Tasks and Critical Path reports as indicators of project health. The goal is to ensure all tasks, critical or not, are completed as planned. These reports identify potential schedule slips and schedule risks, so monitor them often as work proceeds on your project.

Visual Reports

Visual reports take advantage of Project's database by exporting select metadata into Microsoft Excel or Microsoft Visio. These applications have more formatting capability than Project and allow for scaling the report to many different levels of detail.

Figure 9.12: Visual reports use Excel and Visio for charting and graphing.

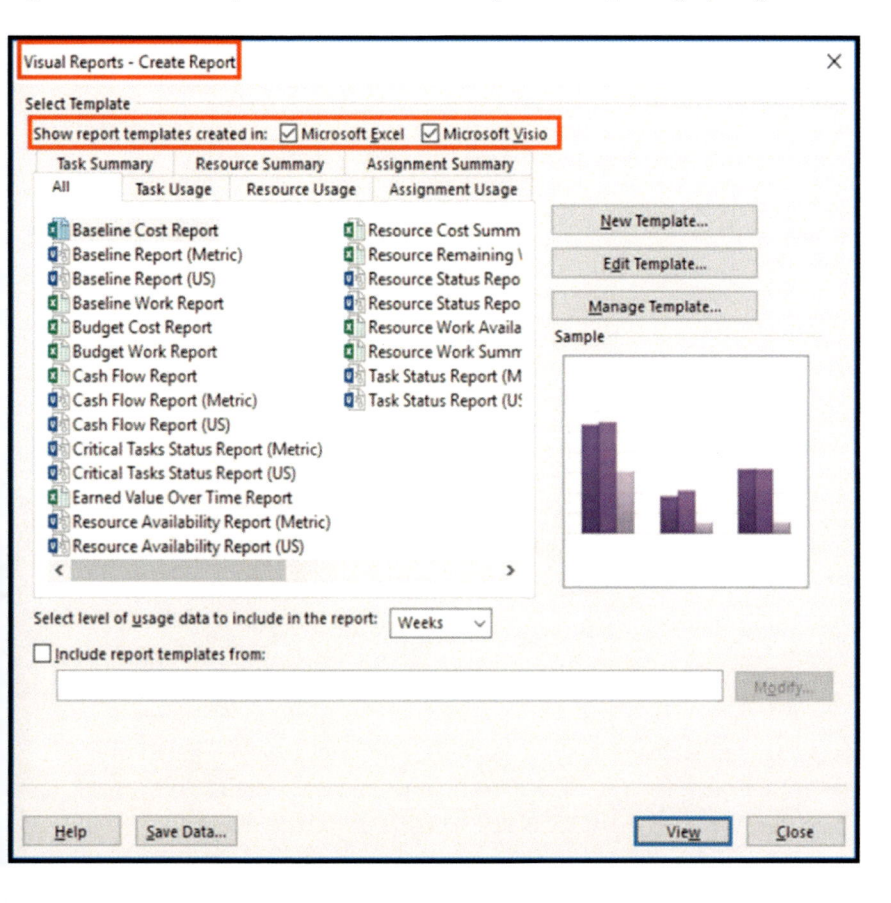

The Cash Flow Report for Excel is an example of a formattable, scalable report. Once the metadata is exported to Excel, a pivot table and pivot chart are created that represent the project's cash flow. The data is shown initially at the Project Summary level, with details contained in the pivot table.

Figure 9.13: The Cash Flow visual report is summarized initially in an Excel pivot chart.

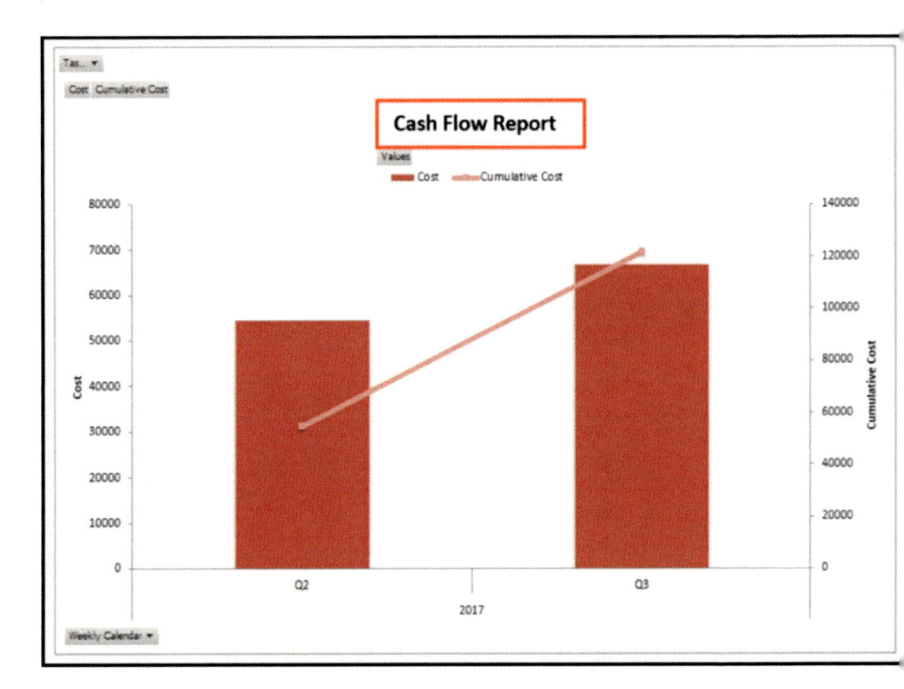

If more detail is desired across periods, the pivot table's date and time control will require a smaller time than represented in the pivot chart. Once the change is made in the pivot table, the change will trigger an update in the pivot table.

Figure 9.14: Date and Time controls in the Excel pivot table.

Figure 9.15: The Excel pivot chart after changing the pivot table time granularity from Quarters to Weeks.

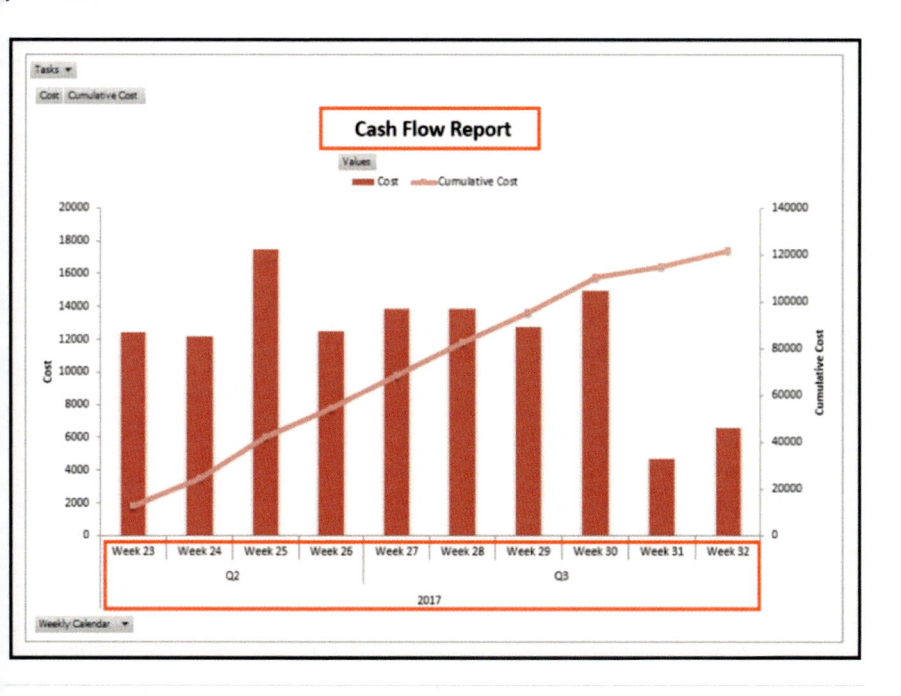

Ch 9: Reporting | 131

✓ **Do This:** Use the Timeline as a focused report. In addition to ease of use, it is intuitive and can be created or modified quickly.

✗ **Don't Do This:** Don't forget that Views can be reports! Details can be rolled up in sheets and charts. You can always show more details when requested.

A Final Thought on Reporting

Microsoft Project was designed to meet the needs of project managers and project leaders. It has amazing breadth and depth, but that does not mean that every feature is called for in every project. Use what is meaningful to the project and leadership team now but leave the "advanced" features and reports until later when the organization is ready for them.

Do's and Don'ts Review

Do's:
- Build your own reports.
- Use the "Slipping Tasks" and "Critical Path" reports as indicators of project health.
- Use the "Timeline" as a focused report.

Don'ts:
- Don't forget that Views can be reports!

Chapter 10: Using Microsoft Project on Agile Based Projects

In this chapter, you will learn:

- The basics of Agile based projects
- The Microsoft Project "Agile Project Management" template
- How to build a "Kanban Board" as a quick scheduling, reporting and tracking tool

Introduction to Agile based projects

Microsoft Project for Windows was introduced to the project management community almost thirty years ago. It was designed to use the methods of the day to schedule in the sequential or "waterfall" technique, which is not iterative. In contrast, Agile project management techniques are iterative. This does not mean they are mutually exclusive techniques! Hybrid project plans contain both sequential and iterative techniques. Microsoft Project can be used in all three scenarios: sequential, iterative and hybrid.

If you are a Project Online or Project Pro for Office 365 user, you have Agile tools available for your use, and they are built right into the interface. No customization is required. You can begin your Agile project management experience immediately. For more information, see https://blogs.technet.microsoft.com/projectsupport/2017/10/30/project-goes-agile/.

Also, see https://support.office.com/en-us/article/use-agile-in-microsoft-project-online-desktop-client-1b9b44d7-fd8e-4b3b-ab94-2b97deb9945b?ui=en-US&rs=en-US&ad=US.

If you use Microsoft Project as a stand-alone product, you still have options for managing projects using Agile techniques. Microsoft has a Project template for Agile projects available. We will investigate this, and also create a new, Agile specific view and report later in this chapter. Before we jump into the template, a few basic concepts and terms need explanation.

Agile project management refers to techniques used to manage iterative projects and which provide the customer product or functionality with every iteration. This promotes customer involvement with project, as with every iteration, the product is re-designed, feedback gathered and improvements made. The most common techniques are Scrum and Kanban. For a more detailed explanation of Agile methodologies, conduct a web search of the terms "Scrum" and "Kanban" using your browser.

Scrum is a method involving the creation of a prioritized list of product components or work items, referred to as the Backlog. Once a Backlog is created, a fixed period of time to perform a fixed amount of work in the backlog is scheduled. The time period is referred to as a Sprint. Each work component is called a Work Item. As a Sprint concludes, the work and product are evaluated and the next Sprint planned. It is an aggressive and fast approach to complex, but rapidly moving projects. The Microsoft Project template is designed around Scrum methodology.

Kanban is a method of delivering project products by focusing on work flow. A single thread of workflow guides each work item through completion stages. Often these are defined as Not Started, In Progress and Completed. No time box is utilized, as in Scrum, as its premise is a continual flow of work and delivery. The technique involves prioritizing the work list, focusing on a manageable amount of work and completing it. For each of the three stages, the work is moved through lanes representing each. The Kanban Board is a framework for managing the work. It looks similar to this:

Figure 10.1: Example of a simple Kanban Board.

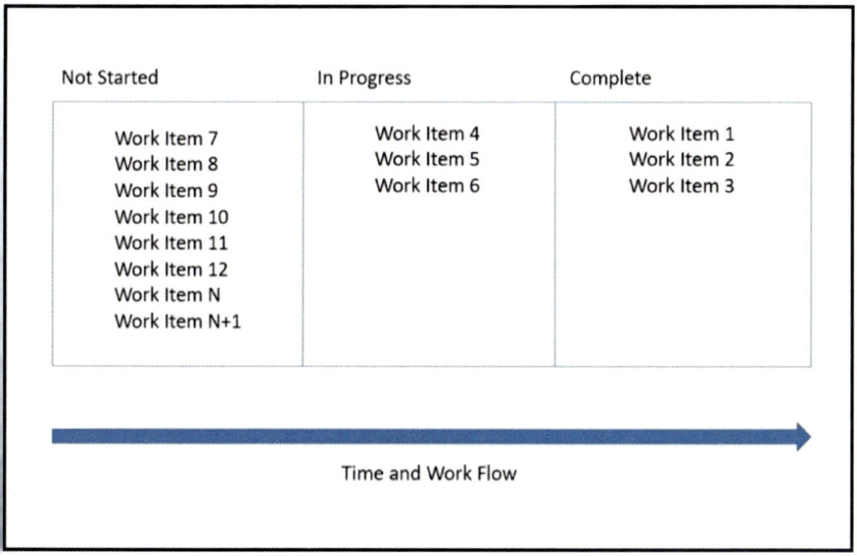

Not Started	In Progress	Complete
Work Item 7　　Work Item 8　　Work Item 9　　Work Item 10　　Work Item 11　　Work Item 12　　Work Item N　　Work Item N+1	Work Item 4　　Work Item 5　　Work Item 6	Work Item 1　　Work Item 2　　Work Item 3

Time and Work Flow →

The Kanban Board can coexist with traditional project task scheduling in Microsoft Project. You will need to perform a small amount of customization, but it's worth the effort. Later in this chapter, we will create Microsoft Project's version of a Kanban Board.

Using the Microsoft Project Agile Project Management Template

This template is Microsoft's attempt to bring Agile into Project 2013. As such, you'll have to make do with the template or customize it to fit your needs. After we examine the template, we will build a simple Kanban Board that can be used in any Project file.

When you first open Project it will take you to the "New" page, which offers you featured templates, as well as for new or previous projects. The Agile template is titled "Agile project management." If it isn't in the New page, click on "Projects" in "suggested Searches:" near the top of your screen.

Figure 10.2: Find the "Agile project management" template.

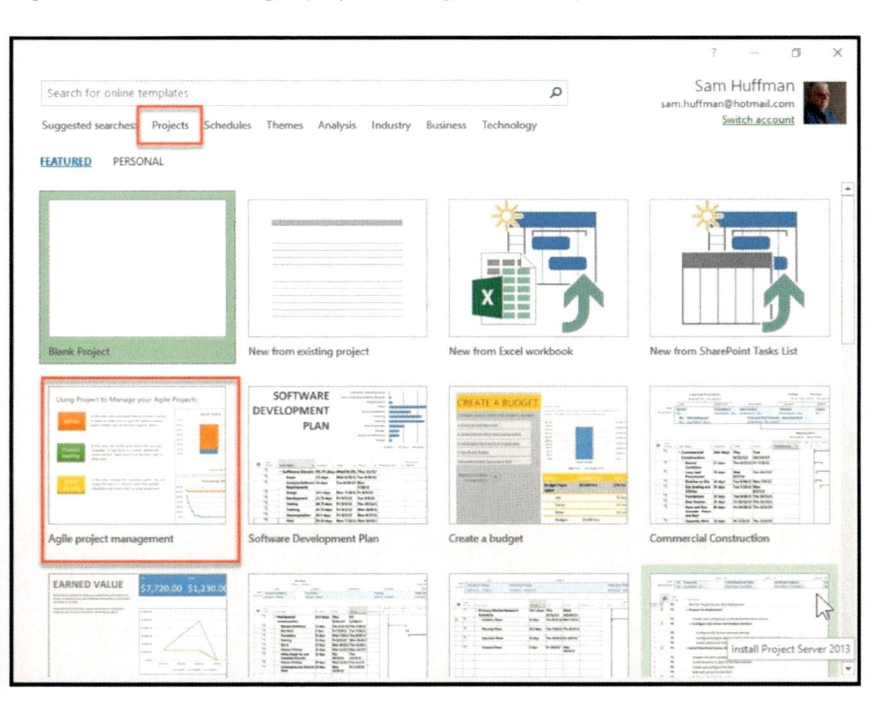

When you find the template, double-click on it. Project will offer you a summary of the template and a "Create" button. Click on the "Create" button to download and open the template.

Figure 10.3: Create the Agile project management template.

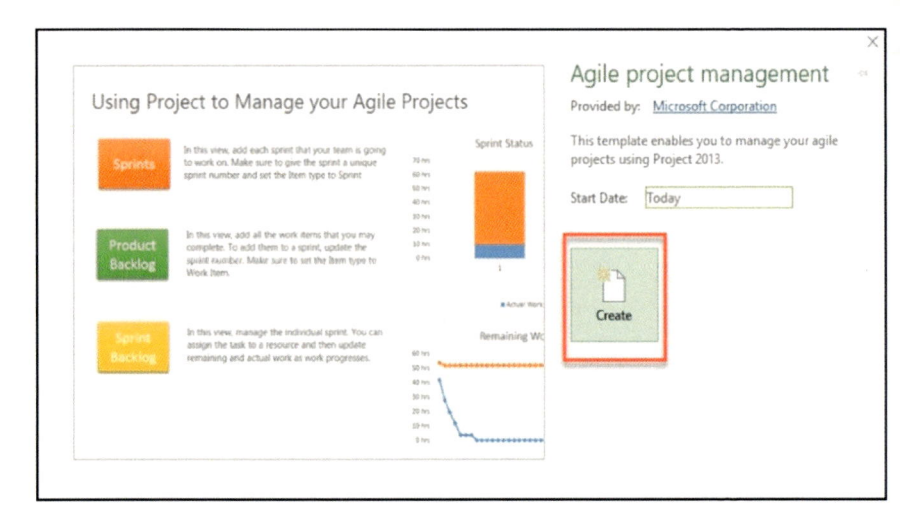

When the template is created, you will be taken to the "SCRUM TEMPLATE INSTRUCTIONS" report. This contains brief, clear instructions on the usage of the template. Each title box to the left of the directions is a shortcut to the specific View used in the template. Click the box of your choice, and it will take you to that specific View. To return to the instructions page, select the "Report" tab, then click on "Custom" and finally on "SCRUM TEMPLATE INSTRUCTIONS." Alternatively, you can place the "View Custom Reports" command in your Quick Access Toolbar, so that you can return to the SCRUM template more quickly. Refer back to *Chapter 1: Harness the Interface* for specifics on adding commands to the Quick Access Toolbar.

> Do This: Read the SCRUM TEMPLATE INSTRUCTIONS report very carefully, including the note beneath the charts. This information will save you frustration when working in the template!

Figure 10.4: Follow directions on the "SCRUM TEMPLATE INSTRUCTIONS" page.

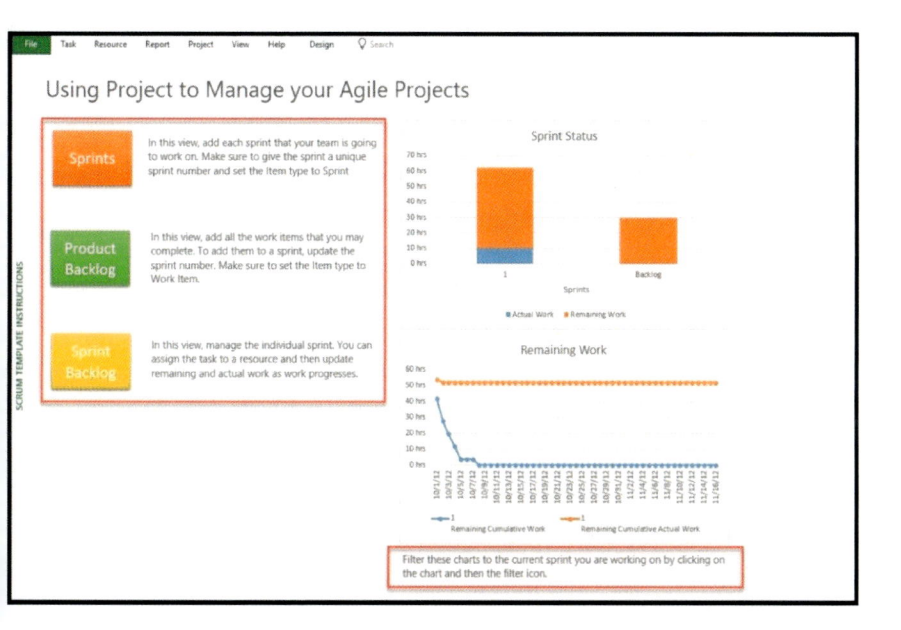

Once you have your basic information into the template, save it as a Project file, rather than a template. You will need to change the project date, adjust the calendar and the other setup steps outlined in *Chapter 2: Setup for Success*.

> Do This: It will take a little practice to use the Scrum template, so embrace and explore it! After you have used it a few times, its simplicity and usefulness may make it one of your favorite Project files.

Create a Kanban Board

Regardless of the project management methodology used, managers and executives often want to know only the basic status of a task or project. Questions like "Has it started?" or "Is it done?" are not calling for a detailed analysis of whether performance is going as planned. They are quick questions demanding a quick answer.

Figure 10.5: The "Kanban Board" clearly shows what is completed, what is in progress and what is not yet started.

#		Task Mode	Task Name	Duration	Start	Finish	Predecessors	Resource Names
1			**Kanban Board: Completed**	9d	6/5/17	6/15/17		
2	✓		Create and negotiate definition documents	3 days	6/5/17	6/7/17		MGT TEAM,PM,QA/QC
3	✓		Create and publish project announcement	5 days	6/8/17	6/14/17	2	MGT TEAM,PM,Printer Paper[3 R
4	✓		Assemble and organize project team	1 day	6/15/17	6/15/17	3	MGT TEAM,PM,QA/QC
5	✓		Control Gate: Planning	0 days	6/15/17	6/15/17	4	
6			**Kanban Board: In Progress**	24.5d	6/16/17	7/20/17		
7			Create tasks and organize per definition documents	1 day	6/16/17	6/16/17	5	PM,QA/QC,Team[300%]
8			Sequence tasks and estimate durations	2.5 days	6/19/17	6/21/17	7	Team[500%]
9			Identify resources and assign to tasks	2.5 days	6/21/17	6/23/17	8	MGT TEAM[200%],PM,QA/QC
10			Level resources and get buy off from management	2.5 days	6/26/17	6/28/17	9	MGT TEAM,PM
11			Control Gate: Begin Work	0 days	6/28/17	6/28/17	10	
13			Conduct work cycle 1	7 days	6/28/17	7/7/17	11	Team[500%]
14			Review for quality 1	1 day	7/7/17	7/10/17	13	QA/QC
15			Conduct work cycle 2	7 days	7/10/17	7/19/17	14	Team[500%]
16			Review for quality 2	1 day	7/19/17	7/20/17	15	QA/QC
			Kanban Board: Not Started	13d	7/20/17	8/8/17		
17			Conduct work cycle 3	7 days	7/20/17	7/31/17	16	Team[500%]
18			Review for quality 3	1 day	7/31/17	8/1/17	16,17	QA/QC
19			Control Gate: Close and Document	0 days	8/1/17	8/1/17	18	
21			Close project books	1 day	8/1/17	8/2/17	19	PM
22			Release resources to management	1 day	8/2/17	8/3/17	21	PM
23			Prepare final report	1 day	8/3/17	8/4/17	22	PM,Printer Paper[2 Reams]

Note that the "Simple Kanban Board" in Figure 10.1 and the "Kanban Board" shown in Project contain similar information. The difference between the two is that one is a formula-driven table in Microsoft Project, while the other is text or even sticky notes in a table. Both work, but by incorporating the Project table, a complete schedule can be managed at any level during the project's life cycle. Additionally, the project manager can choose the power of Project's features or the simplicity of Kanban to manage the project.

The "Kanban" technique was originally imagined by Toyota to capture workflow. It can have more than the three states used in this example depending on need and preference. Conduct a web search on "Kanban" to get more details on this excellent technique.

In order to create a "Kanban Board" in Microsoft Project, a small amount of customization must be done. A custom field, a custom view and a custom group all interrelate and give you this simple and effective tool.

The custom field should be an unused text field. This is critical! You will overwrite existing data if you choose a text field already in use.

Here are the steps to create a "Kanban Board:"

1. Select the "Project" tab, and then click on "Custom Fields."
2. Select an open text field, then click on the "Rename…" button to give the text field a unique name. I chose "Kanban Board" for consistency.

Figure 10.6: Select an open text field for customization.

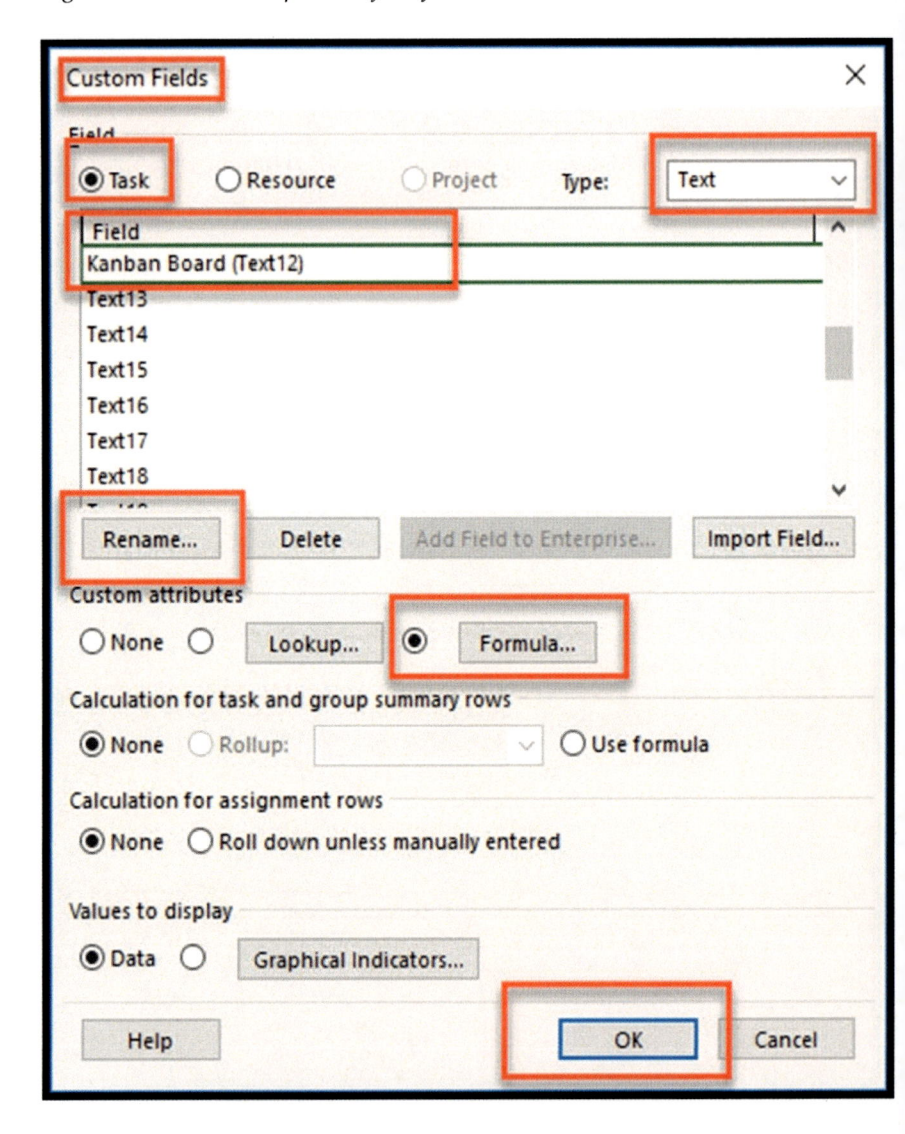

3. Click on the "Formula…" button to enter the formula that will control the Kanban Board.

4. Enter the formula as shown in the figure below. You may have to adjust the formula for your specific system. For example, some systems require a single quote rather than the double quote.

Figure 10.7: Create the formula that powers the Kanban Board field.

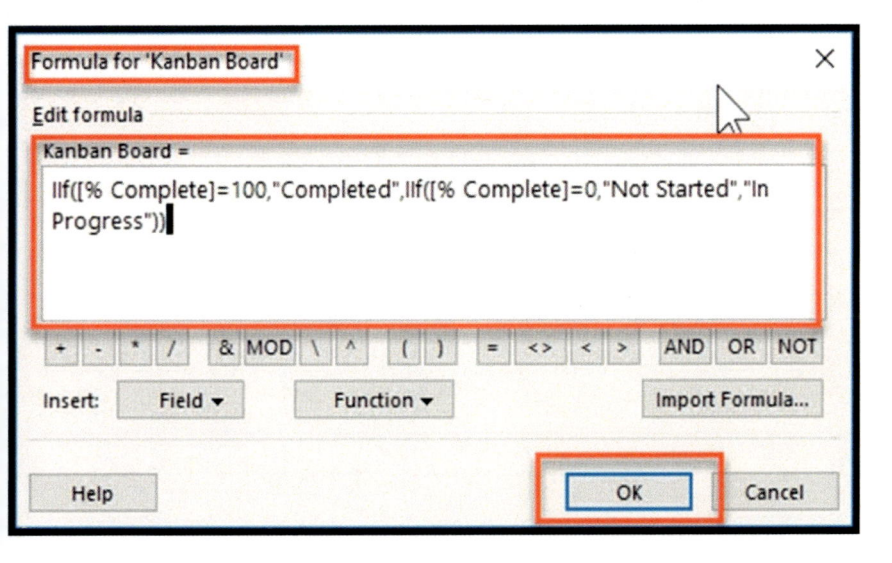

```
Formula for 'Kanban Board'
Edit formula
Kanban Board =
IIf([% Complete]=100,"Completed",IIf([% Complete]=0,"Not Started","In Progress"))
```

5. Once the formula is entered, select "OK." You now have the custom field needed for the "Kanban Board." You do not need to display this field. It will calculate the state of completion for each task for use in groups, views and reports.

6. Prepare a Group to organize the project by completion state. Select the "View" tab. Then from the "Group by:" dropdown list, click on "New Group By..." and enter the information below into the Group Definition dialog.

7. Click on the "Save" button to save your new custom group.

Figure 10.8: Create a Group to format the "Kanban Board."

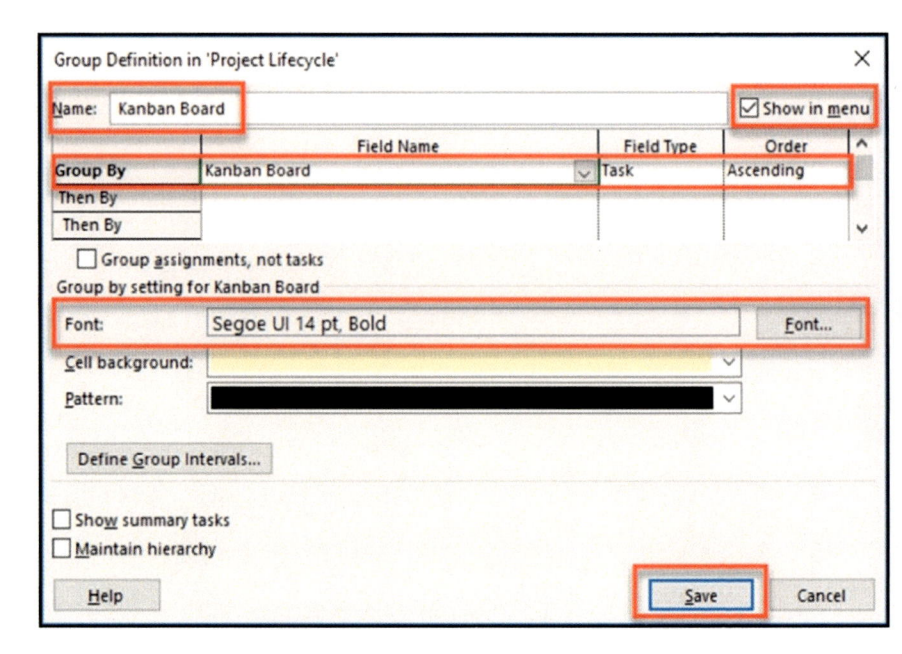

8. Select the "View" tab and from the list of "Task Views" click on "More Views…" and "New…" to create a new View to use the new field and group.

9. Enter the information below into the new View Definition dialog. Don't forget to click on "OK."

Figure 10.9: Create the "Kanban Board" view.

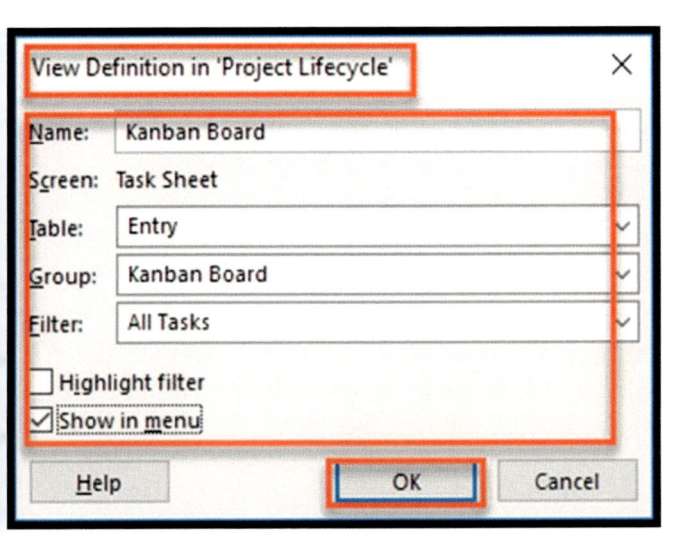

View Definition in 'Project Lifecycle'	
Name:	Kanban Board
Screen:	Task Sheet
Table:	Entry
Group:	Kanban Board
Filter:	All Tasks
☐ Highlight filter	
☑ Show in menu	
Help	OK Cancel

10. Finally, test your new view by selecting the "Task" tab, then selecting your "Kanban Board" from the list of custom task views. It should look similar to the figure below.

Figure 10.10: The result of careful customization is your "Kanban Board."

		Task Mode	Task Name	Duration	Start	Finish	Predecessors	Resource Names
			Kanban Board: Completed	**9d**	**6/5/17**	**6/15/17**		
2	✓		Create and negotiate definition documents	3 days	6/5/17	6/7/17		MGT TEAM,PM,QA/QC
3	✓		Create and publish project announcement	5 days	6/8/17	6/14/17	2	MGT TEAM,PM,Printer Paper[3 F
4	✓		Assemble and organize project team	1 day	6/15/17	6/15/17	3	MGT TEAM,PM,QA/QC
5	✓		Control Gate: Planning	0 days	6/15/17	6/15/17	4	
			Kanban Board: In Progress	**24.5d**	**6/16/17**	**7/20/17**		
7			Create tasks and organize per definition documents	1 day	6/16/17	6/16/17	5	PM,QA/QC,Team[300%]
8			Sequence tasks and estimate durations	2.5 days	6/19/17	6/21/17	7	Team[500%]
9			Identify resources and assign to tasks	2.5 days	6/21/17	6/23/17	8	MGT TEAM[200%],PM,QA/QC
10			Level resources and get buy off from management	2.5 days	6/26/17	6/28/17	9	MGT TEAM,PM
11			Control Gate: Begin Work	0 days	6/28/17	6/28/17	10	
13			Conduct work cycle 1	7 days	6/28/17	7/7/17	11	Team[500%]
14			Review for quality 1	1 day	7/7/17	7/10/17	13	QA/QC
15			Conduct work cycle 2	7 days	7/10/17	7/19/17	14	Team[500%]
16			Review for quality 2	1 day	7/19/17	7/20/17	15	QA/QC
			Kanban Board: Not Started	**13d**	**7/20/17**	**8/8/17**		
17			Conduct work cycle 3	7 days	7/20/17	7/31/17	16	Team[500%]
18			Review for quality 3	1 day	7/31/17	8/1/17	16,17	QA/QC
19			Control Gate: Close and Document	0 days	8/1/17	8/1/17	18	
21			Close project books	1 day	8/1/17	8/2/17	19	PM
22			Release resources to management	1 day	8/2/17	8/3/17	21	PM
23			Prepare final report	1 day	8/3/17	8/4/17	22	PM,Printer Paper[2 Reams]

If you wish to create a report based on the view, simply create a new table report using the "New Report" wizard in the "Report" tab, and apply the "Kanban Board" group in the report's Field List. Show all tasks in the outline level, and you have the complete tool in Project.

To migrate your work into another file or into the Global Template, refer to *"Appendix A: The Organizer."*

Do's and Don'ts Review

Do's:
- Read the SCRUM TEMPLATE INSTRUCTIONS report very carefully, including the note beneath the charts. Use the "Slipping Tasks" and "Critical Path" reports as indicators of project health.
- It will take a little practice to use the Scrum template, so embrace and explore it!

Chapter 11: Agile Project Management using Microsoft Project Online Desktop Client

This chapter has been contributed by PPM Expert, Erik van Hurck.

In this chapter you will learn:

- The difference between the One-Time purchase and Subscription versions of Microsoft Project
- About the new fields views and Ribbon that come with the Agile functionality
- Best practices for using the Agile toolset

The Project Online Desktop Client

In previous chapters, Sam has covered functionality that's possible with any Microsoft Project Professional application. This chapter contains information that is only applicable for those who have the Project Online Desktop Client.

If you are not sure if you have the Project Online Desktop Client, navigate to File, and then click Account. Locate the version and build number of the application in the top right corner. If you have Project Online Desktop Client, you will see the following image:

Figure 11.1: Subscription Product License.

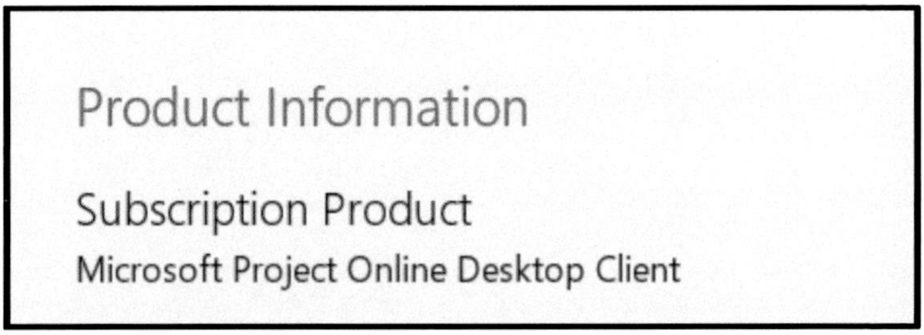

Product Information

Subscription Product
Microsoft Project Online Desktop Client

Another easy way to make sure you have the subscription version of Microsoft Project is to navigate to File, and then to New. Confirm that you see Sprints Project or Scrum and Kanban Project templates on the New menu.

Figure 11.2: New menu displays Springs Project or Scrum and Kanban Project options.

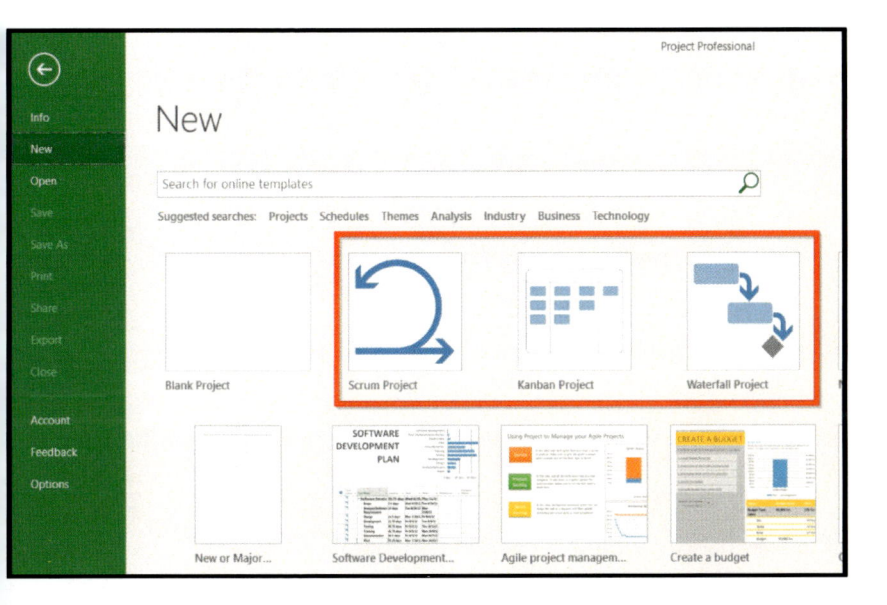

Or

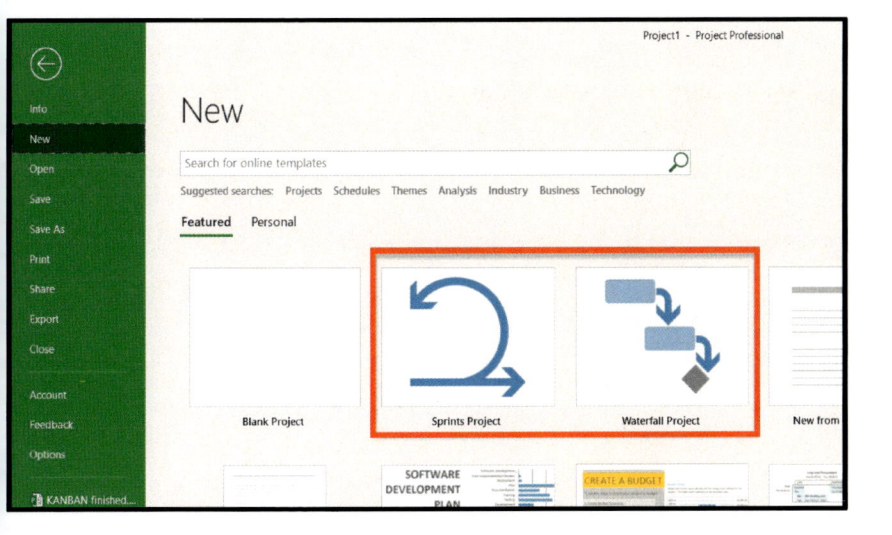

Ch. 11: Agile Using Project Online Desktop | 153

The Cloud Approach

The Project Online Desktop Client application version is included with licenses for Project Online Professional and Project Online Premium, and it is part of Microsoft's cloud strategy. The Office 365 Suite includes, among other software, Word, Excel and PowerPoint "in the cloud."

All subscription versions of Microsoft's applications contain new and improved functionality. New features will be added much faster than the versions that are not subscription based. In fact, there are some situations within the Microsoft family of products where a subscription version contains content that is not a part of other versions of the product.

In the case of Microsoft Project, the subscription version contains added functionality that is not found in other versions. One such functionality is the whole new Agile module. At the time of this writing, the newest version of Microsoft Project's desktop application is Microsoft Project 2019, but it does not contain the Agile module, because that is only available on the subscription version, Project Online Desktop Client.

> Do This: Get the subscription version of Microsoft Project to make use of the new Agile module described in this chapter.

Connection with Enterprise Solutions

Both the "Professional" and Online subscription versions of Microsoft Project (i.e. Microsoft Project Professional 2019 and Project Online Desktop Client) can connect to the Enterprise solution called Project Server or Project Online.

Enterprise solutions contain many more capabilities than single Project files do. To learn more, refer to MPUG's website, where there are various articles describing Enterprise in detail.

Continuous Updates

When subscribed to a cloud based subscription like Project Online Desktop Client, users get a continuously improving application.

You can determine what version of Project Online Desktop Client you are running by navigating to File, and then clicking on Account. On the right side, there is a section called "About Project," and the version number is visible there.

Figure 11.3: About Project displays the version number.

Note that either one or two Agile templates are available in the New menu. This is related to the version of the Online application you have (Version 1904 and higher will have just one Agile template in the New menu).

The images and functionality described in this chapter are related to version 1904. There will be additions to this chapter if new functionality is added that significantly changes the content.

Creating Your First Sprints Project

Now that you have established that you have a Project Online Desktop Client version of the application, let's have a closer look at the new content that comes with the Agile functionality. We can start by creating a Sprints project:

1. Open Projcct.
2. Click on New from the File menu.
3. Click on Sprints Project.

Figure 11.4: Click on Sprints Project from the New menu.

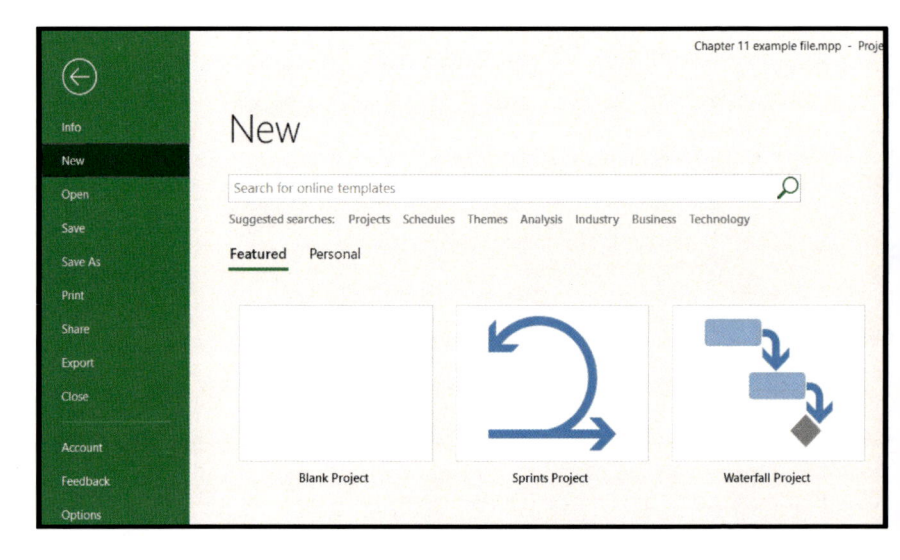

You will be presented with a completely new view called the "Sprint Planning Board."

Figure 11.5: Sprint Planning Board view.

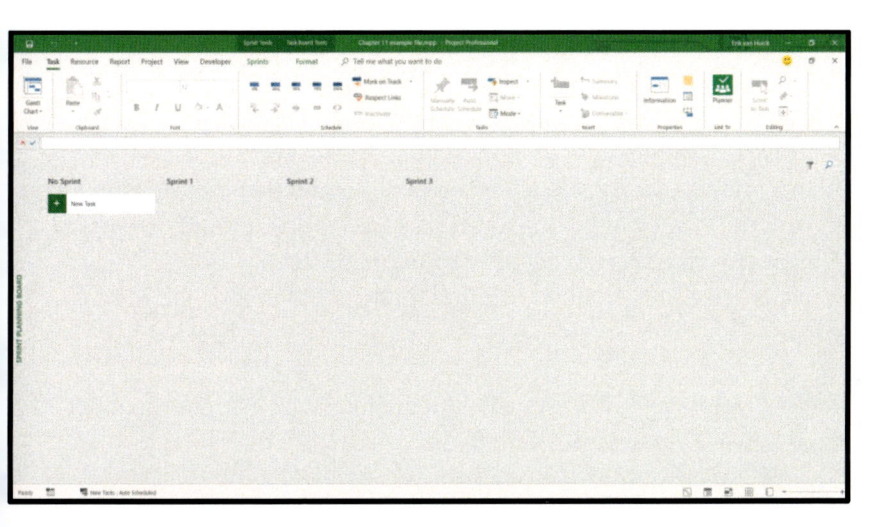

This view lacks the table and Gantt chart visual representations of the schedule. Instead, there are four sections with the titles "No Sprint," "Sprint 1," "Sprint 2" and "Sprint 3." There is also a "New Task" button that you can interact with. If you look at the Ribbon, you will see two new menu's called "Sprint Tools" and "Task Board Tools."

New Fields

Before creating any tasks, let's first have a look at the new fields views and Ribbon. There are six new fields related to Agile functionality found in this version of Project. They are listed by column name below.

Show on Board: Indicates tasks to be included in Board views and Task Board reports. The type of field is a Yes/No Flag.

Sprint: Adds the task to a sprint. Look up table field type.

Board Status: Indicates the status of the task from the Task Board view. Look up table field type.

Sprint ID: Number (formula) field type showing the Sprint ID.

Sprint State: Date field to view the start date of the sprint the task is in.

Sprint Finish: Date field to view the finish date of the sprint the task is in.

Both new fields that contain a look up table (Sprint and Board Status) are a special type. Instead of having a direct influence on the content, these fields present two special methods for filling the look up table. We will cover this more fully in the Using the Agile Toolset section later in this chapter.

> Do This: Hover over a column name to get a brief description of what the field is meant to do.

> Don't Do This: Create custom fields with the same name as the default fields.

New Views

There are various new views related to the Agile functionality in Project Online Desktop Client. The fields each view contains "out of the box" are listed below.

Sprint Planning Sheet: Fields included by default are ID (locked), Indicators, Sprint, Name, Work, Board Status, Resource Names, Task Summary Name, Deadline, Show on Board and Add New Column.

Sprint Planning Board: A task that is created only shows up with the task name and assigned resources. When first navigating to this view there are four columns: No Sprint, Sprint 1, Sprint 2 and Sprint 3.

Current Sprint Board: A mix between the Task Board view and the Sprint Planning Board. This shows a Task Board, but filters on only the current sprint activities.

Current Sprint Sheet: Fields included by defaults are ID (locked), Indicators, Sprint, Name, Work, Board Status, Resource Names, Task Summary Name, Deadline and Add New Column.

Task Board Sheet: Fields included are ID (locked), Indicators, Sprint, Name, Work, Board Status, Resource Names, Task Summary Name, Deadline and Add New Column.

Task Board: A task that is created only shows up with the task name and assigned resources. When first navigating to this view there are four columns: Not Started, Next up, In progress and Done. To the right, there is an Add New Column option, and below each column we have the option to include a percentage complete value.

Backlog Board: A board similar to the Task Board.

Backlog Sheet: Fields included are ID (locked), Indicators, Name, Work, Board Status, Resource Names, Task Summary, Deadline and Add New Column.

All Agile Tasks: Fields included are ID (locked), Indicators, Sprint, Name, Work, Board Status, Resource Names, Task Summary Name, Deadline and Add New Column.

Note that the last three views listed come from an earlier version of the Agile functionality where there was a difference between Scrum and Kanban. Since then, Microsoft decided to go for one Agile solution. The Backlog views and Agile views were filtered, but are no longer a part of the newer versions.

The Board views don't have columns in the sense that we are used to. The views are changed to "Boards," which look a lot like the previous Kanban or Sprint board. All tasks that have "Show on Board" set to "Yes" are represented as a card containing relevant data.

Don't Do This: Don't use the default views Backlog Board, Backlog Sheet or All Agile Tasks, as they are not working as intended anymore.

Creating your own version of a Board view is straight forward. Select Task, Gantt Chart and then More Views... or New...

From the menu shown in the figure below, you can select the screen type "Task Board."

Figure 11.6: Create your own Board view.

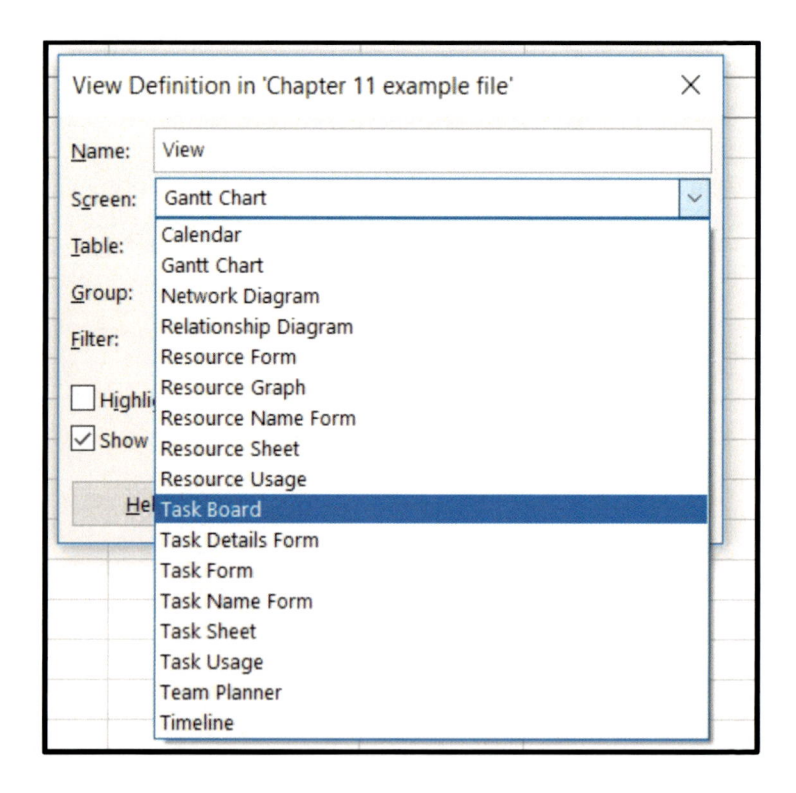

When you have Task Board selected, add a Filter if desired. Note this is the only option available.

Figure 11.7: Add a Filter is the only available option.

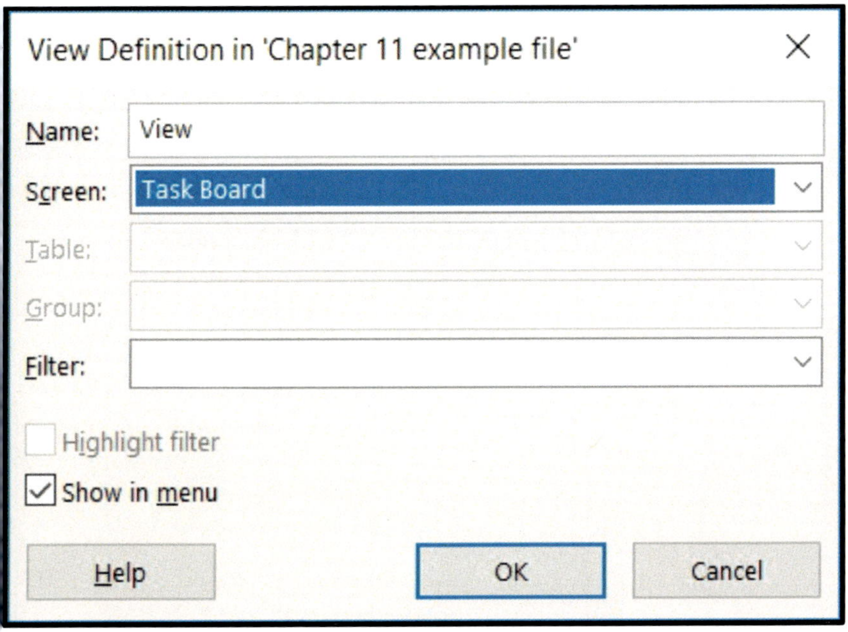

View Definition in 'Chapter 11 example file'	
Name:	View
Screen:	Task Board
Table:	
Group:	
Filter:	

☐ Highlight filter
☑ Show in menu

[Help] [OK] [Cancel]

New Ribbon

Within the new Agile module, there are new Ribbon tabs created specifically to work with the new views and added functionality. The three new Tabs are listed below, as well as associated Groups and the actions that can be performed.

Table 11.1: New tabs with group name and actions listed.

Tab Name	Group Name	Actions
Sprint Tools / Sprints	Views	Task Board, Planning and Sprint
	Sprints	Manage
Task Board Tools / Format	View	Sheet
	Customize	Customize Cards and Show % Complete Mapping
Task Sheet Tools / Format	Format	Text Styles and Gridlines
	Columns	Insert Column, Align Left, Center, Align Right, Wrap Text, Column Settings, Custom Fields
	Show/Hide	Outline Number, Project Summary Task and Summary Tasks

There are also other sections on the Ribbon that now include Agile functionality. On the Project Tab, for example, you will now find the Manage Sprints action, and on the View Tab, you will find the Task Board action. Additionally, the Report Tab now includes a Task Boards report action. All the actions can be included in a personalized Ribbon Tab.

> ❌ Don't Do This: Don't remove the default actions on the default Tabs.

Using the Agile Toolset

When setting up Sprint Duration and dates, remember that the Sprint Planning Board is the default view that you start with when creating a project using the Sprints template (see images 11.4 and 11.5).

To change the number of sprints and when these sprints start, navigate to the Sprint Tool Sprints Tab.

Figure 11.8: Sprint Tool Sprints Tab.

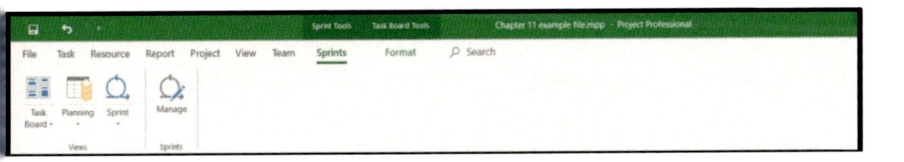

Click on "Manage" to open up the Manage Sprints menu.

Figure 11.9: Manage Sprints.

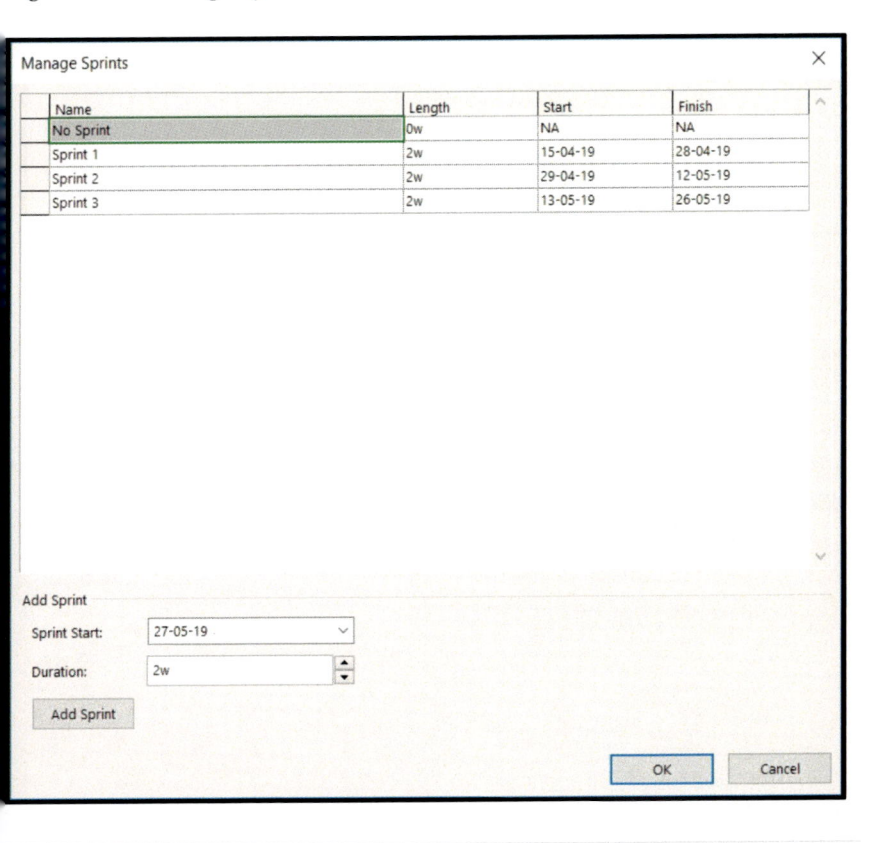

Name	Length	Start	Finish
No Sprint	0w	NA	NA
Sprint 1	2w	15-04-19	28-04-19
Sprint 2	2w	29-04-19	12-05-19
Sprint 3	2w	13-05-19	26-05-19

Add Sprint

Sprint Start: 27-05-19

Duration: 2w

Add Sprint

Ch. 11: Agile Using Project Online Desktop | 163

The Manage Sprints menu provides you with the option to create additional Sprints or remove them. Add a Sprint by clicking on the button, Add Sprint, which is located on the lower left side of the menu. A new Sprint will be created on the date that is specified in the "Sprint Start:" date field. The new Sprint will have the duration specified in the "Duration:" field.

Note: Changing the duration with the buttons creates Elapsed Weeks instead of normal working weeks. It is best to keep a single Sprint Duration until the completion of the "project." Another thing to be aware of is that if you delete a sprint, the tasks that were associated with that sprint will be associated with the value "No Sprint" instead.

Figure 11.10: Deleting a Sprint confirmation dialog box.

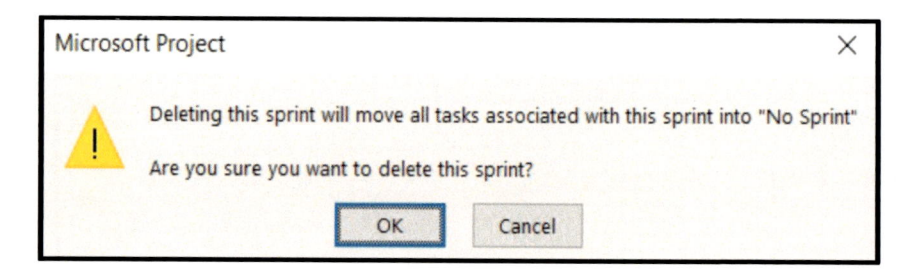

There are two things to note when it comes to Sprint dates:

The tasks you create within a sprint do not change their start or finish dates. Remember this when navigating between Gantt chart or Usage views and/or the actual Sprint board and Agile related views. It would be wise to set the start date of the task as soon as work commences. This will create a more accurate resource utilization for the current (and upcoming, if you plan ahead) Sprints.

✅ Do This: Update start dates for tasks to reflect the Sprint start dates. This is done by right-clicking on a task from within the Task Board or Sprint Planning Board and selecting "Information." The start date field is located on the General tab.

Secondly, Sprint dates actually behave similarly to deadlines in a schedule. Any changes to Gantt chart related dates doesn't change the dates within the Sprint dates mentioned in the Manage Sprints menu. This means that changing the duration or start dates of tasks doesn't reflect back to the Sprint dates. Note, changing the project start date doesn't reflect on the Sprint dates either! It is best practice to create a whole list of Sprints with dates when a project starts.

✅ Do This: Make sure that the Project Start date is the actual date you want to start. Creating sprints (including the first three) will reflect that start date. Changes afterwards will only be reflected on the Sprint dates.

Setting Up the Task Board

The Task Board is similar to a standard Kanban board. It contains the columns "Not Started," "Next up," "In Progress" and "Done." Additional columns can be added by clicking on the "Add New Column" option to the right of "Done." Some organizations prefer to have an additional column between "In Progress" and "Done" that simulates "Evaluate," "UAT" or just "Test." In any case, know that your organization's situation can be simulated as required on the Task Board.

Figure 11.11: Default Task Board set up.

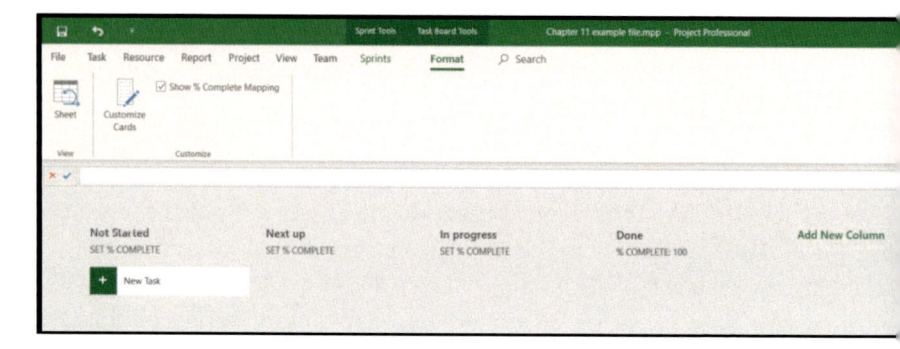

> **Do This:** Do create additional columns if your business process requires it.

Underneath the column titles there is a "% complete mapping" value that can be set for any of the columns. By default, the "Done" column is set to 100% complete. This mapping is reflected on the Gantt charts and will change the percentage complete on activities as they progress towards the "Done" status / column.

You can, for instance, state that "In Progress" items are 50% done by adding this value to the % complete mapping value associated with the "In Progress" column.

Figure 11.12: In progress value.

```
In progress
% COMPLETE: 50
```

If you don't want to see the mapping values, de-select the checkbox for "Show % complete mapping." This setting can be found on the Customize tab in the Task Board Tools Format Ribbon.

✓ **Do This:** Change the percentage complete values for Sprint tasks by using the Task Board view. This will reflect on the Gantt views, as well.

✗ **Don't Do This:** Don't update the percentage complete values for Sprint tasks by using the Gantt view, as they will not reflect back to the Task Board view.

Setting Up Card Values

There's not much to the default task card. There's the task name and associated resources, if you have them. In the case of a large group of resources on a single card, the card will grow to include them.

Figure 11.13: Task card grows to include resources, as needed.

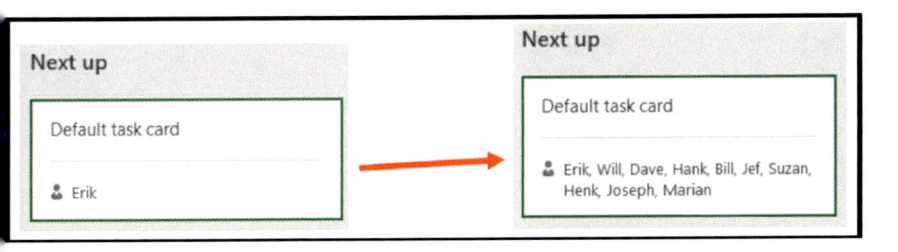

If a task is completed, a small checkmark will be displayed. This can happen either when the task is marked 100% complete, or if it is moved to a "Done" column that has 100% completed mapping.

Figure 11.14: Checkmark displays when task is completed.

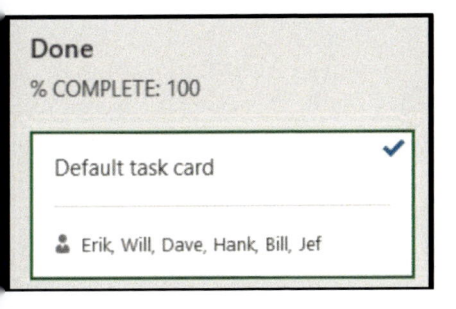

Microsoft has now made it possible to "Customize Cards." The control for this feature can be found on the Ribbon in the Task Board Tools Customized group.

Figure 11.15: Customize Task Board Cards.

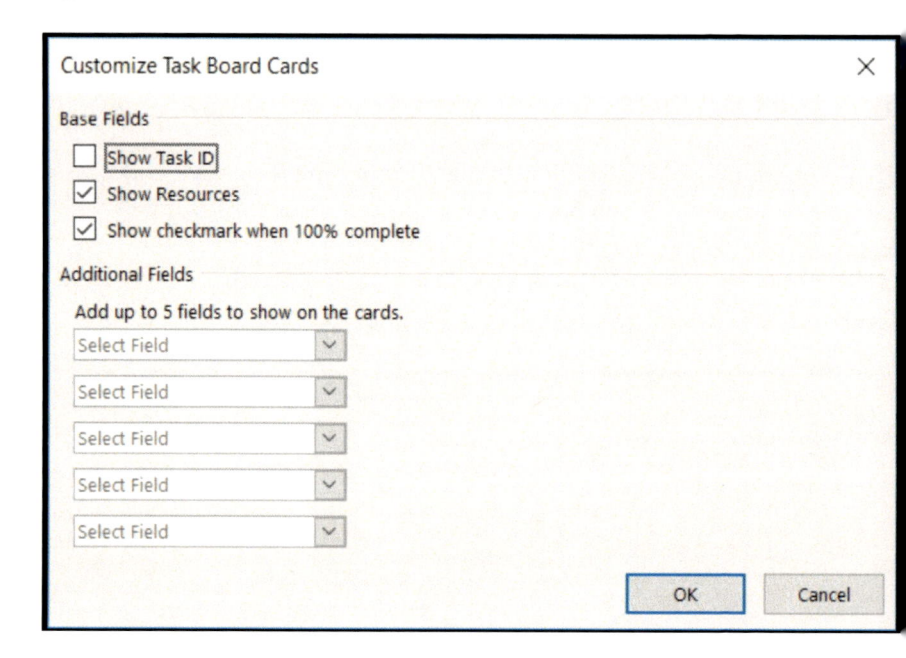

In the Customize Task Board Cards dialog box, there are "Base Fields" and "Additional Fields." The Base Fields section gives you the option to Show Task ID, Show Resources and Show checkmark when 100% complete. When unchecked, the corresponding item is not shown on the Task Board Card.

In the Additional Fields section, Project provides you with the option to add up to five fields on the card. The following figure shows a card set to display Work and Cost fields.

Figure 11.16: Work and Cost fields displayed on Task Board card.

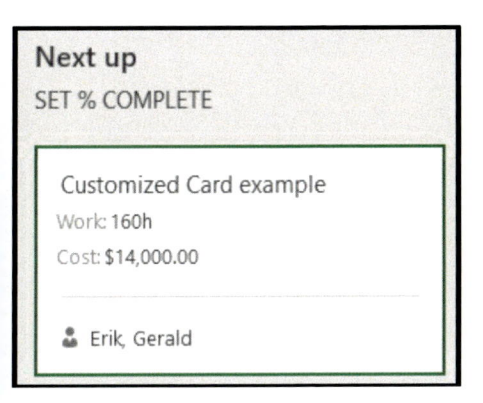

```
Next up
SET % COMPLETE
  ┌─────────────────────────────────┐
  │ Customized Card example         │
  │ Work: 160h                      │
  │ Cost: $14,000.00                │
  │ ─────────────────────────────── │
  │ 👤 Erik, Gerald                 │
  └─────────────────────────────────┘
```

The example shown is not very Agile though, is it? An Agile mindset would call for the use of Sprint Tokens and User Story association fields.

✅ <u>Do This:</u> Do create Agile fields, and add them to the cards for a better Agile experience.

Sprint Token: Adding a custom number field gives you an easy way to manage Sprint Tokens. Make sure the field has been set to roll-up a summary of the tokens. You can debate if Sprint Tokens don't actually equal Hours work spent, but that goes beyond the goal of this text.

User Story Association: Create a custom look up text field to associate Task Board Cards with User Stories. Filling the look up table list will ensure easy reporting on specific user stories.

Figure 11.17: Display additional custom fields for a more Agile way of working.

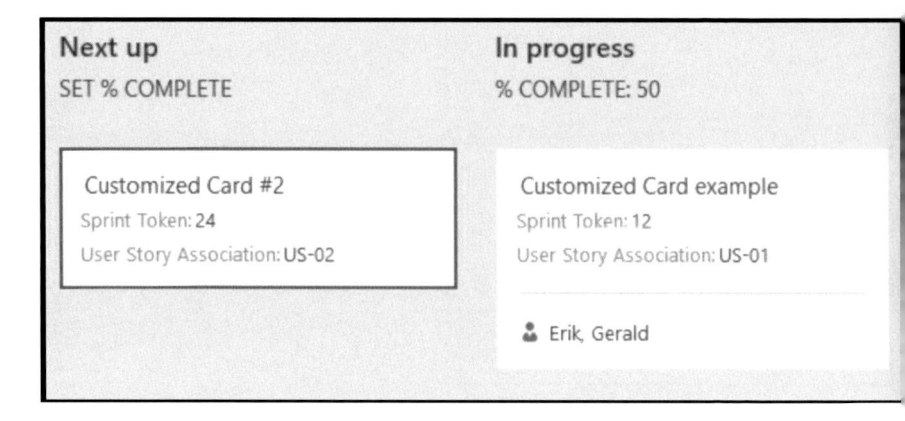

Next up	In progress
SET % COMPLETE	% COMPLETE: 50
Customized Card #2 Sprint Token: 24 User Story Association: US-02	**Customized Card example** Sprint Token: 12 User Story Association: US-01 👤 Erik, Gerald

Once created, both of these fields can also be added to the Task Board Sheet or Sprint Planning Sheet to further enhance the user experience.

Figure 11.18: Consider adding custom Agile fields to the Sprint Planning Sheet.

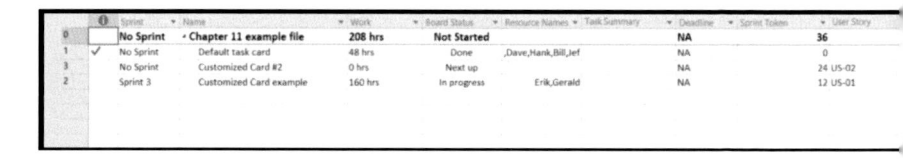

	Sprint	Name	Work	Board Status	Resource Names	Task Summary	Deadline	Sprint Token	User Story
0	No Sprint	Chapter 11 example file	208 hrs	Not Started			NA	36	
1	✓ No Sprint	Default task card	48 hrs	Done	Dave,Hank,Bill,Jef		NA	0	
3	No Sprint	Customized Card #2	0 hrs	Next up			NA	24	US-02
2	Sprint 3	Customized Card example	160 hrs	In progress	Erik,Gerald		NA	12	US-01

An Alternative Solution for User Stories

A good solution for referencing User Stories can be that of hijacking the Summary tasks. If a Schedule is fully Agile, Summary tasks do not serve any purpose other than ones we give them.

Linking User Stories to Summary tasks provides an advantage when using the Filter option in the Board Views. The filter gives you the option to sort by Resources or Summary tasks, and when the Summary tasks have the User Stories as their names, the filter tool becomes easier than ever to use.

The Task Summary name is already mentioned on most of the Sheets associated with Agile functionality, so it makes perfect sense to apply this solution.

> **Don't Do This:** Don't retain the Summary Tasks' default setting of "Show on Board" as this setting as it will cause tasks to show up as cards themselves.

The following figures show this methodology applied on the Sprint Planning Sheet and full Spring Planning Board, as well filtered views.

Figure 11.19: Sprint Planning Sheet with User Stories as Summary tasks.

Sprint	Name	Work	Board Status	Resource Names	Task Summary Name	Deadline	Sprint Token	Show on Board
No Sprint	Chapter 11 example file	0 hrs	Not Started			NA	100	Yes
No Sprint	As a User of MS Project I want to plan in a Agile way	0 hrs	Not Started			NA	50	No
Sprint 1	Purchase Project Online Desktop Client	0 hrs	Not Started		As a User of MS Project I want to plan in a Agile way	NA	5	Yes
Sprint 2	Activate the Office Insider track	0 hrs	Not Started		As a User of MS Project I want to plan in a Agile way	NA	20	Yes
Sprint 2	Create Custom Fields	0 hrs	Not Started		As a User of MS Project I want to plan in a Agile way	NA	25	Yes
No Sprint	As a Manager I want to have great Agile reports	0 hrs	Not Started			NA	50	No
Sprint 3	Address the issues with the Out of The Box reports	0 hrs	Not Started		As a Manager I want to have great Agile reports	NA	15	Yes
Sprint 3	Change color template to match that of the company	0 hrs	Not Started		As a Manager I want to have great Agile reports	NA	35	Yes

Figure 11.20: The full Sprint Planning Board.

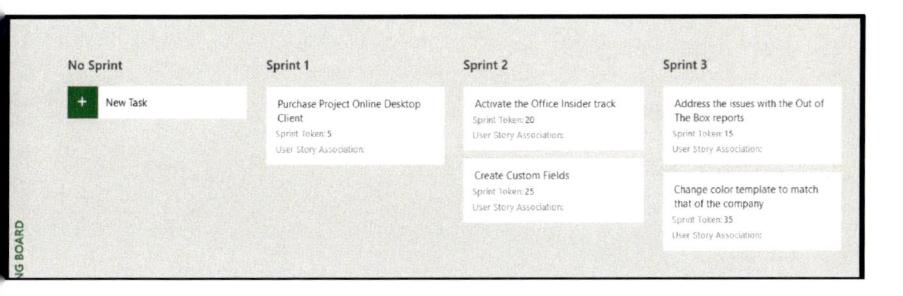

No Sprint	Sprint 1	Sprint 2	Sprint 3
+ New Task	Purchase Project Online Desktop Client Sprint Token: 5 User Story Association:	Activate the Office Insider track Sprint Token: 20 User Story Association:	Address the issues with the Out of The Box reports Sprint Token: 15 User Story Association:
		Create Custom Fields Sprint Token: 25 User Story Association:	Change color template to match that of the company Sprint Token: 35 User Story Association:

Figure 11.21: Filtered view of one User Story.

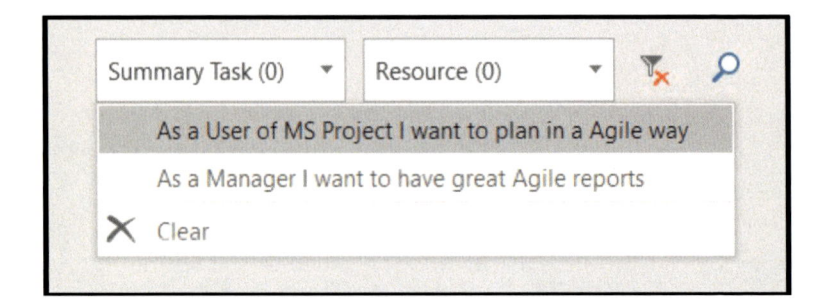

Figure 11.22: Filtered view of the Sprint Planning Board.

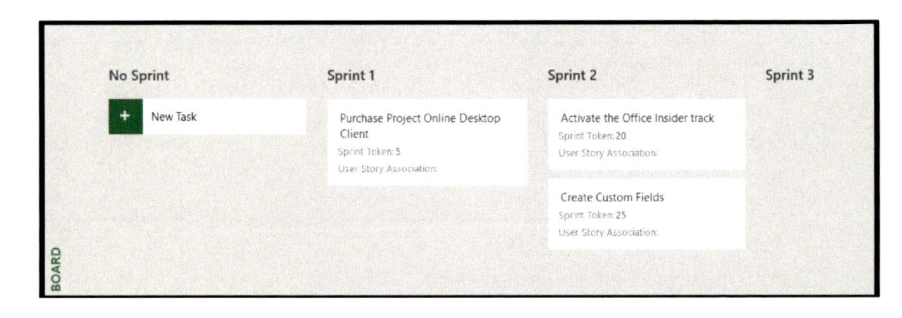

Do's and Don'ts Review

Do's:
- Get the subscription version of Microsoft Project to make use of the new Agile module described in this chapter.
- Hover over a column name to get a brief description of what the field is meant to do.
- Update start dates for tasks to reflect the Sprint start dates. This is done by right-clicking on a task from within the Task Board or Sprint Planning Board and selecting "Information." The start date field is located on the General tab.
- Make sure that the Project Start date is the actual date you want to start. Creating sprints (including the first three) will reflect that start date. Changes afterwards will only be reflected on the Sprint dates.
- Do create additional columns if your business process requires it.
- Change the percentage complete values for Sprint tasks by using the Task Board view. This will reflect on the Gantt views, as well.
- Do create Agile fields, and add them to the cards for a better Agile experience.

Don'ts:
- Create custom fields with the same name as the default fields.
- Don't use the default views Backlog Board, Backlog Sheet or All Agile Tasks, as they are not working as intended anymore.
- Don't remove the default actions on the default Tabs.
- Don't update the percentage complete values for Sprint tasks by using the Gantt view, as they will not reflect back to the Task Board view.
- Don't retain the Summary Tasks' default setting of "Show on Board" as this setting as it will cause tasks to show up as cards themselves.

Appendix A: The Organizer

When Microsoft Project is started, a file named Global.mpt is opened and then read by the program. This file is a template that contains the set-up configuration required to display your data and calculate work and task schedules correctly.

The Organizer is opened by clicking on the File tab, then Info. Finally, click on Organizer.

Figure A.1: Getting to the Organizer.

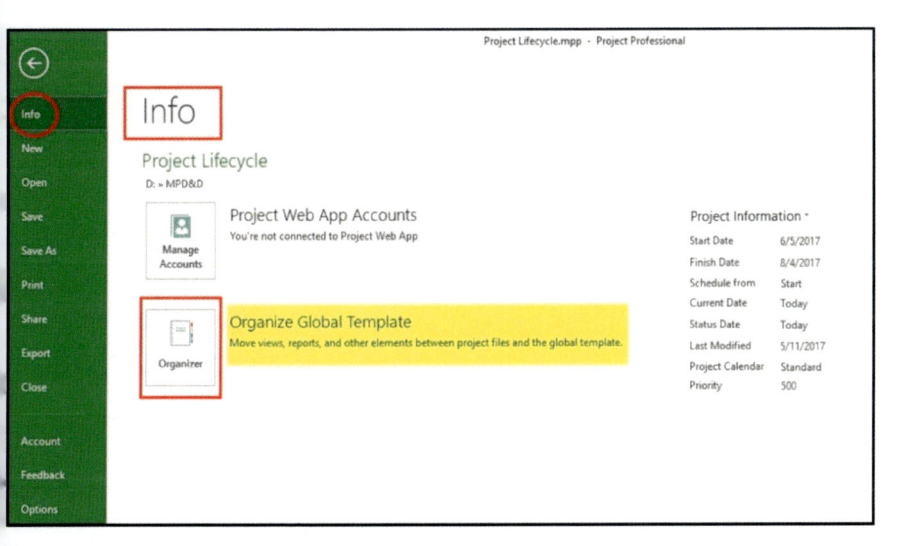

Once the Organizer opens, you can use it to copy components between files. The tabbed interface identifies the components that are interchangeable between Project files. You work hard to create Reports and Calendars. The Organizer allows you to leverage that work. It is a real time saver when you need it.

Figure A.2: The Organizer contains interchangeable file components to save file development time.

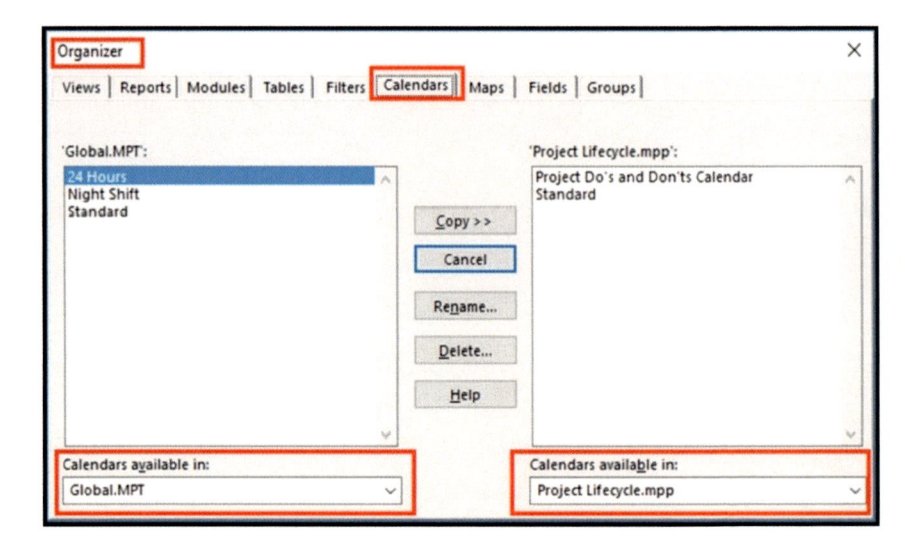

How You Might Use Organizer

One example of the Organizer's use is in the selection of calendars. If a custom calendar were created and copied to Global.mpt, it could then be copied to any file *not* containing the custom calendar. The only requirement is that Global.mpt and the file needing the custom calendar be opened and listed in a file list at the bottom of the Organizer dialog.

Figure A.3: Copy the "Project Do's and Don'ts Calendar" from its Project file by opening the Organizer and selecting the Calendars tab. Select the Project Do's and Don'ts Calendar and then click on the Copy button to copy it into Global.mpt.

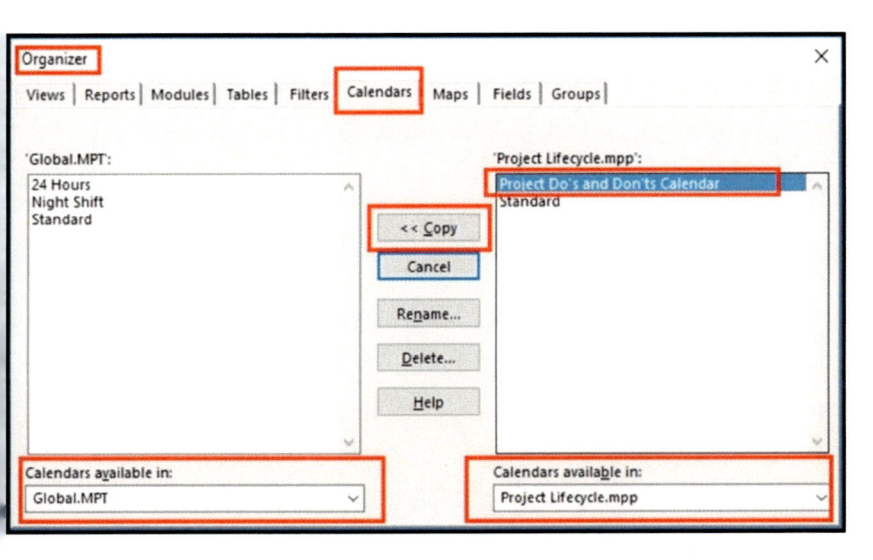

Figure A.4: The Organizer showing that the custom calendar "Project Do's and Don'ts Calendar" has been copied to the Global template. The custom calendar is now available in any open Project file.

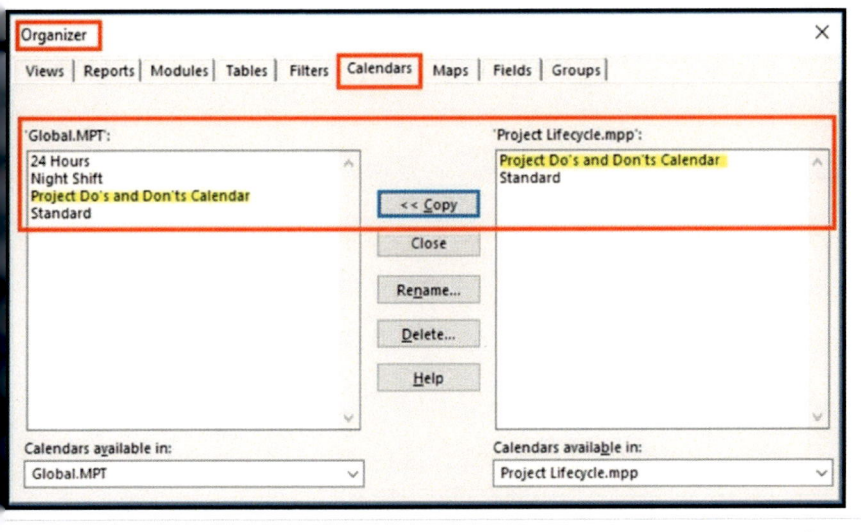

Figure A.5: Project Information opened in a new file showing the availability of the "Project Do's and Don'ts Calendar." By selecting it and clicking OK, the Standard calendar is replaced by the Project Do's and Don'ts Calendar.

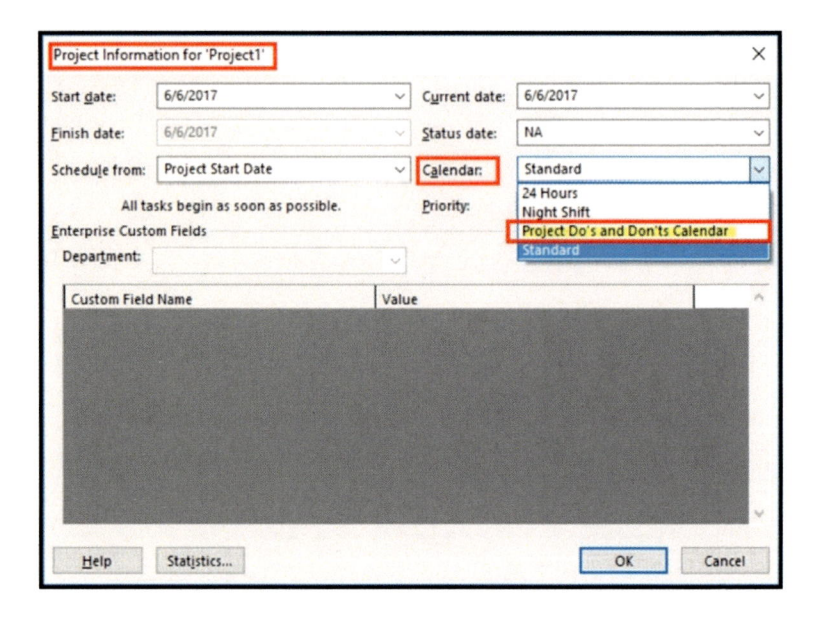

Appendix B: Using Microsoft Project Help

When you have a tough problem in Project, there is a good chance that the solution is in Project Help. Its catalog of answers is searchable, dynamic and very interactive. You can enter Help from just about anywhere in the interface. Here are a few tips and tricks to help you get the full benefit of Help.

Use the F1 key on your keyboard! Pressing F1 in any view will call up the Project Help home page. This page offers you a search tool and a categorized list of Help topics.

Figure B.1: The Help home page.

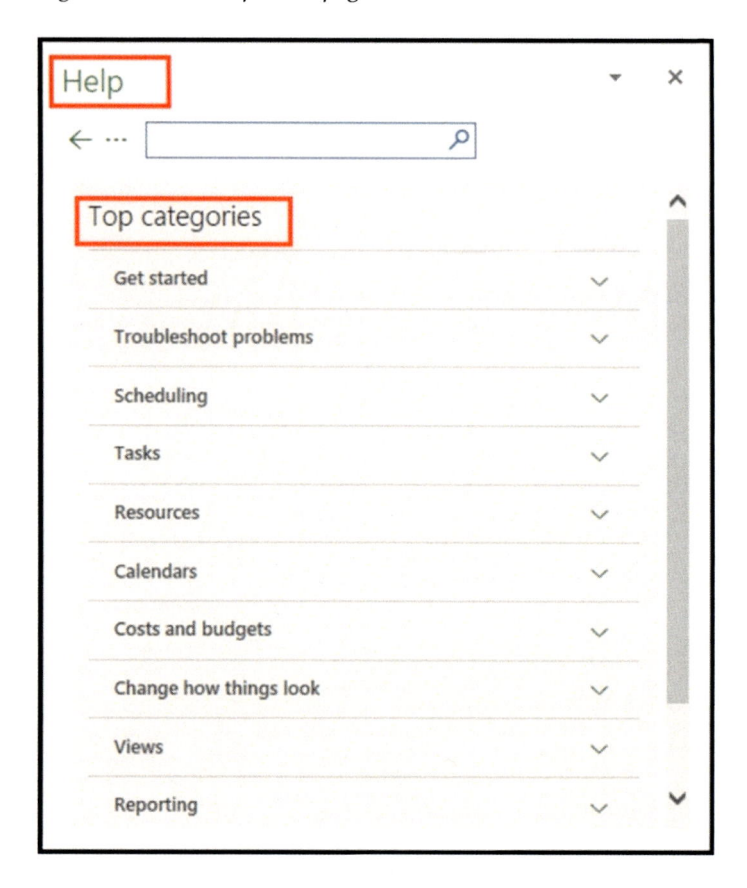

Help	
Top categories	
Get started	
Troubleshoot problems	
Scheduling	
Tasks	
Resources	
Calendars	
Costs and budgets	
Change how things look	
Views	
Reporting	

Use F1 from within a dialog box such as Change Working Time. Pressing F1 while in this kind of dialog, for example, will bring up online Office Help informing you about how to work with calendars in Project. Additionally, the article will have the versions of Project that the answer will pertain to. In the case of Change Working Time, the answer contains explanations, links to other relevant topics and how to fine tune calendars in your project.

Figure B.2: An Office Help support topic called from within a dialog box.

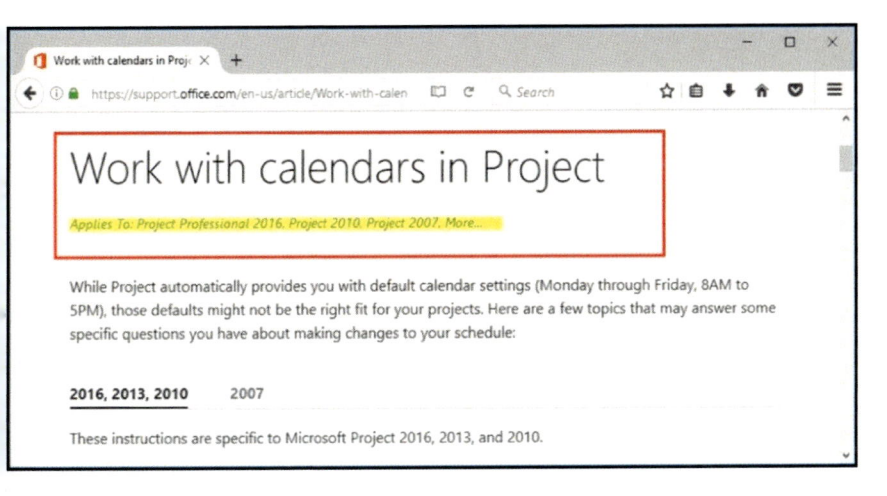

Use Tell me what you want to do! New in Project 2016 is an additional entry point to the Help topics. When you click on "Tell me what you want to do," a search box and a short list of topics is offered. Tell me what you want to do is found to the right of the Format tab.

Figure B.3: "Tell me what you want to do" is an entry point into Help.

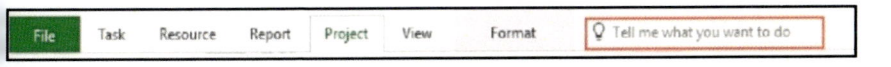

Typing a subject or topic in the search box will create a link to the Help topic at the bottom of the list. A difficult topic like task types are especially good ones to search on and learn about. Clicking on these topic links can result in unexpected "jewels" of information such as the task type revision table found in Figure B.6.

Figure B.4: Searching and selecting Help from "Tell me what you want to do."

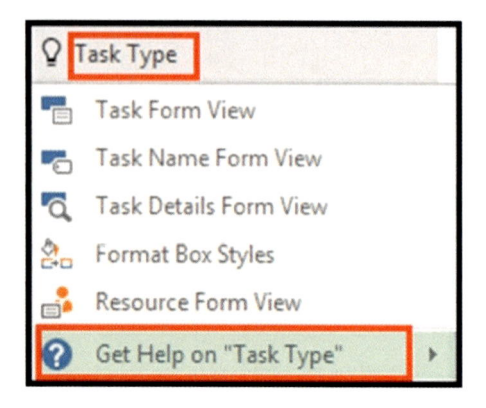

💡 Task Type
🗐 Task Form View
🗐 Task Name Form View
🔍 Task Details Form View
Format Box Styles
Resource Form View
❓ Get Help on "Task Type"   ▸

Figure B.5: The Get Help topic produces another short list of relevant topics.

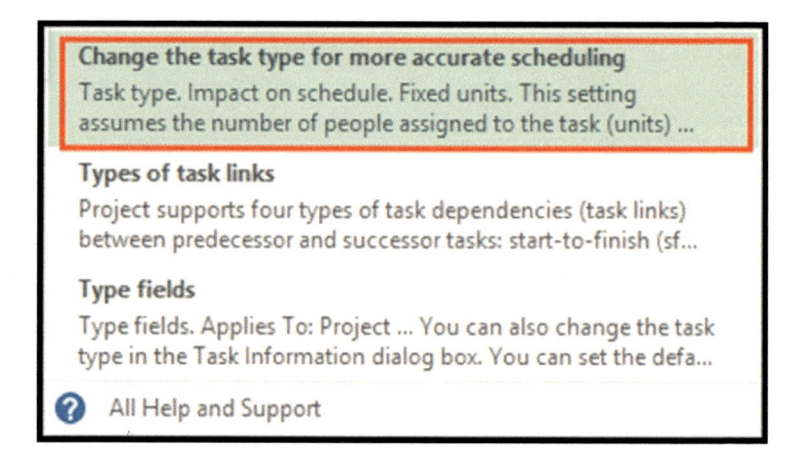

Change the task type for more accurate scheduling
Task type. Impact on schedule. Fixed units. This setting assumes the number of people assigned to the task (units) ...

Types of task links
Project supports four types of task dependencies (task links) between predecessor and successor tasks: start-to-finish (sf...

Type fields
Type fields. Applies To: Project ... You can also change the task type in the Task Information dialog box. You can set the defa...

❓ All Help and Support

Figure B.6: A Task type revision table from Help.

In a	If you revise units	If you revise duration	If you revise work
Fixed units task	Duration is recalculated.	Work is recalculated.	Duration is recalculated.
Fixed work task	Duration is recalculated.	Units are recalculated.	Duration is recalculated.
Fixed duration task	Work is recalculated.	Work is recalculated.	Units are recalculated.

Here are some useful Help topics to search on:

- Keyboard shortcuts
- Add holidays and vacation days
- Create subtasks and summary tasks
- Add lead or lag time to a task
- Manage costs
- Assign people to work on tasks
- Enter costs for resources
- Level resource assignments
- Save a Project file as PDF

Appendix C: Steps to Create a Project File

Set up the project to receive your data entry (Chapter 2):

1. Select a project start date.
2. Create or adjust the project calendar.
3. Choose a project scheduling method.

Create the task list (Chapter 3):

1. Turn on outlining tools.
2. Identify summary tasks and enter them.
3. Identify sub-tasks for each summary task. Enter them into the list and indent accordingly.
4. Consider which task scheduling method to use.

Enter duration estimates (Chapter 4):

1. Estimate each task based on these formulas:
   a. Work = units x duration
   b. Units = work / duration
   c. Duration = work / units
2. Choose the task type for each task.

Sequence the sub-tasks in the task list. (Chapter 5):

1. Carefully examine each task and determine if it has a predecessor or successor. Create the relationship.
2. Enter or modify relationships on the successor tasks only.
3. Enter lead and lag as required.
4. Doublecheck relationships to ensure no summary task has a predecessor or successor.

Create, assign and level resources. (Chapter 6):

1. If you are going to use resources, enter them on the Resource Sheet. If you will be tracking Actual Work and Actual Cost, define their resource type, their calendars, their Max Units and their rate.
2. Assign resources to their respective tasks.
3. Look for overallocations.
4. Identify the severity and cause of overallocations. Compare Max Units to Peak Units and look for imbalances in their values.
5. Doublecheck to ensure no resources are assigned to summary tasks.
6. Level the project.

Baseline the project (Chapter 7):

1. Choose a baseline method: entire project or partial.
2. If partially baselining, remember to set the baseline to roll up to summary tasks.

Track the project. (Chapter 8):

1. Establish a tracking interval that will support your reporting interval.
2. When entering progress, always enter the actual start date first. When finished, enter the actual finish date.
3. Use the Move command when rescheduling tasks.
4. Use the Tracking Gantt and Tracking table to enter tracking data.
5. Use the tracking fields for tracking accuracy: %Complete, %Work Complete, Actual Work and Remaining Work.

Report on the project progress until completion (Chapter 9):

1. Stick to your tracking and reporting schedules.
2. Use Project Statistics as a quick summary of the project. It will help you predict finish dates and communicate delays.
3. Identify stakeholder "Hot buttons" and strive to report on them in every report cycle.

Index

%Complete 110, 113

A

Actual cost 109
Actual duration 110
Actual finish 109
Actual start 109
Actual work 109-114, 120, 124
Add new column 10-12, 78
Adjust lead and lag 61
Agile project management 135-148
  template 138-140
Agile toolset 162
Assigning a cost resource 72-76
Assigning a material resource .. 72-76
Assigning a work resource 72-76

B

Backlog 137
Baselines 96
  clearing 96, 100
  setting 15, 96-98
  tasks 99-101
  variance 101-105
Bottom-up design 44

C

Calendars used by Project .. 23-26, 27-30, 151-154
Calendars, creating 27
  hierarchy 26
  modifying 27
  resource 25, 31
  setting 27
  shift 25
Cash flow report 125
Change working time 27-30, 156
Changing scheduling mode 34
Charts 5-6, 13, 120-123
Clear leveling values 82, 92

Cloud approach 154
Combination view 8, 82-83, 90, 114
Command groups 4, 6
Cost resource 45, 70, 75-76
Critical task report 126-127

D

Dashboards reports 120
Dependencies 63-65
Duration ... 10, 13, 15, 22, 31-32, 38-40, 48-54, 83, 99, 101, 110, 120, 125

F

Field list 122
Fields 10-12
Finish to finish (F-F) 60
Finish to start (F-S) 59
Fixed duration 53
Fixed units 52-53
Fixed work 53
Forms 14
Full screen view 8

G

Gantt chart 8-10, 13, 15, 62-63, 73, 75-76
Global.mpt 151-153
Graphs 13, 122

I

In progress report 126

K

Kanban 136-137
Kanban Board 137-138
  create the "Kanban Board" view 141-148

L

Lag 60-63, 159
Late tasks report 127
Lead 60-63, 159
Leveling 78-92
  algorithm 79

available slack 88-89
buttons 91-92
calculations............................. 80-82
clearing leveling values 82, 92
creating splits in remaining work
  89
individual assignments 88
manually scheduled tasks 89
methods 79
options 79-80, 92
overallocations............ 77, 79-81, 84
proposed booking type............... 89
range settings............................... 83
resources................................. 78-89
settings............................. 80-87, 92
tactics 79
tools... 89
Leveling Gantt 82-83, 89-92

M

Mark on track.......................... 111-112
Master project 68-69
Material resource.................. 69, 72-74
Max units........................... 71-72, 77-78
Metadata 5, 128-129
Milestones................................... 48-49
MPUG discussion forums 69

O

Organizer, the 151-153
Outlining................................ 38-40, 42
Overallocated resources report.... 124

P

Peak 77-78
Predecessor................................... 58-65
Project baselines............................. 96
Project calendar.......................... 23-31
Project help 155-159
Project information dialog. 22, 24, 31-32, 104
Project file

creating 161-162
Project interface............................2-18
Project leveling algorithm............... 79
Project Online Desktop Client...... 152
Project start date...............22-23, 31-32
Project statistics 104-105
Project summary task 38-44
Project TechCenter........................... 69

Q

Quick access toolbar 2-4, 6-7, 9-10

R

Remaining duration...................... 110
Remaining work......110-111, 114-115, 117, 120-124
Report tab 5, 120
Report tools.................................... 122
Reporting 118-132
Rescheduling work........................ 115
Resource, material 69, 72-74
Resource allocation view 90
Resource calendars 25-27, 31
Resource cost overview report..... 126
Resource graph......... 14, 17, 78, 82, 90
Resource leveling (see "Leveling resources")
Resource pool 68-69
Resource schedules, evaluating 76
Resource sheet..9, 16, 25, 70-72, 77-78
Resource types 69-72
  assigning 72-76
Resources 5, 9, 14, 16, 23-25, 31, 33, 38, 44, 68-78
  creating 70
  reports 123-125
Ribbon 2-4, 6

S

Scheduling 28, 31-34
  automatic 33
  choosing a method 31

from project finish date 32
from project start date 31
manual... 33
method..................................... 31-34
Scheduling mode, changing......34, 49
Scrum 137
SCRUM TEMPLATE
   INSTRUCTIONS 140-141
Sequencing tasks 5, 15, 31-34, 54, 58-65, 68
Set baseline warning 101
Sheets 6, 9-10
Shift calendars................................. 25
Slipping tasks report 127-128
Sprints 156
Start date...................................... 22-23
Start to finish (S-F)........................... 60
Start to start (S-S)............................. 60
Sub-tasks... 38
Subprojects 68
Subscription based software 155
Successor........................... 58-61, 53-65
Summary tasks ...38-44, 48, 64, 72, 93, 100, 146
Summary Task..38-44, 48, 64, 72, 100, 120, 146

T

Tables 9
Tabs 4-6
Tasks .. 13-17, 22-23, 31-34, 38-45, 48-54, 58-65, 72-73, 75, 77-79, 81, 83, 85-86, 88-89, 96, 98-101, 108, 110, 167-172
   assigning resources 72-76
   auto scheduled............................ 34
   durations 48-54
   manually scheduled................... 33
   oganizing............................... 38-45
   outlining 38
   reports................................ 126-128
   sequencing... 5, 15, 31-34, 54, 58-65
   tab 5
Task calendars............................25-27
Task durations............ 13, 31-32, 48-54
Task insertion points 43
Task list 38-44, 58-65
Task path....................................64-65
Task priority 85-86
Task type revision table 157-159
Task types 51-54
"Tell me what you want to do" ....
   157-158
Template for Agile.................. 138-140
Timeline 13, 17-18, 58, 98, 117, 132
Top-down design............................ 41
Total work...................................... 110
Toyota.. 142
Tracking definitions 109
Tracking Gantt 15-16, 98-99, 102
Tracking methods 111
Tracking your project............. 108-118

U

Units, fixed 52-53
Units of resource.........................49-52
Upate tasks 112

V

Variance 98-99, 101-105
   viewing 102
Views.. 4-9, 15, 70-71, 73, 76-77, 82-83, 89-90, 98, 102, 104, 114, 132, 146
   combination........... 8, 82-83, 90, 114
   recommended 15-18
   tracking Gantt 15, 98, 102, 114
Visual reports 5, 128-129

W

Work...49-53
Work resource 31, 69, 74
Working time 23-24, 27-30, 48, 156

Made in the USA
Monee, IL
12 September 2019